Dan
With Gratitude
in Peace
Jonathan
SF
2016

TOWARDS A
NUCLEAR WEAPON
FREE WORLD

TOWARDS A NUCLEAR WEAPON FREE WORLD

Manpreet Sethi

Editor

Introduction by
Air Commodore Jasjit Singh, AVSM, VrC, VM, IAF (Retd)

KW Publishers Pvt Ltd

New Delhi

in association with

Centre for Air Power Studies

New Delhi

The **Centre for Air Power Studies** is an independent, non-profit, academic research institution established in 2002 under a registered Trust to undertake and promote policy-related research, study and discussion on the trends and developments in defence and military issues, especially air power and the aerospace arena, for civil and military purposes. Its publications seek to expand and deepen the understanding of defence, military power, air power and aerospace issues without necessarily reflecting the views of any institution or individuals except those of the authors.

Jasjit Singh
Director
Centre for Air Power Studies
Subroto Park
New Delhi 110010

Tele: (91-11) 25699131
E-mail: office@aerospaceindia.org

Published in India by
Kalpana Shukla
KW Publishers Pvt Ltd
NEW DELHI: 4676/21, First Floor, Ansari Road, Daryaganj, New Delhi 110002
E-mail: knowledgeworld@vsnl.net
MUMBAI: 15 Jay Kay Industrial Estate, Linking Road Extn., Santacruz (W), Mumbai 400054

ISBN 10: 81-87966-83-1
ISBN 13: 978-81-87966-83-8

Typeset by Dharana, New Delhi and Printed at Rajkamal Electric Press. New Delhi

Contents

List of Contributors

JASJIT SINGH is Founder Director, Centre for Strategic and International Studies, India. He currently heads an independent think-tank, Centre for Air Power Studies, New Delhi. He was Director, Institute for Defence Studies and Analyses for 14 crucial years (1987-2001), the period during which India underwent seminal transformation. He was awarded the Padma Bhushan by the President of India for outstanding service to the nation in the field of defence and strategic affairs.

MANMOHAN SINGH is Prime Minister of India

SERGIO DE QUEIROZ DUARTE is High Representative for Disarmament at the Under Secretary General level. A career diplomat, he holds the rank of Ambassador in the Brazilian Foreign Service, where he has served for 48 years.

PRANAB MUKHERJEE is External Affairs Minister, Government of India

MANI SHANKAR AIYAR is Minister for Panchayati Raj and Development of the North-Eastern Region of India.

MOHAMMED HAMID ANSARI is Vice President of India.

JONATHAN GRANOFF is President, Global Security Institute and Co-chair of the Blue Ribbon Task Force on Nuclear Nonproliferation and, Fellow, World Academy of Arts and Sciences.

RAJESH RAJAGOPALAN is Professor of International Politics at Jawaharlal Nehru University, New Delhi and author of *Second Strike: Arguments about Nuclear War in South Asia* (2005) and *Fighting Like a Guerrilla: The Indian Army and Counterinsurgency* (2008).

DOUGLAS ROCHE, O.C. is Chairman, Middle Powers Initiative. Former Senator and Canadian Ambassador for Disarmament, he was Chairman of the

UN Disarmament Committee at the 43rd General Assembly in 1988.

SVERRE LODGAARD is former Director, Norwegian Institute of International Affairs (NUPI). He is a Member of the Norwegian Government's Advisory Board on Security and Disarmament Affairs and author of *Nuclear Proliferation and International Security* (2007).

GEORGE PERKOVICH is Vice President for Global Security and Economic Development Studies and Director of the Nonproliferation Programme at Carnegie Endowment for International Peace. Author of *India's Nuclear Bomb* and co-author of *Universal Compliance: A Strategy for Nuclear Security*.

LI CHANG-HE, a former Chinese diplomat, is Vice President, China's Arms Control and Disarmament Association (CACDA).

IVAN SAFRANCHUK is Advisor to the World Security Institute and lecturer, Moscow State Institute of International Relations. He is also engaged in private consultancy on security and foreign policy issues.

MANPREET SETHI is Senior Fellow, Centre for Strategic and International Studies, India and Fellow, International Relations, *Centre de Sciences Humaines*, New Delhi. She also heads the Nuclear Security Project at the Centre for Air Power Studies, New Delhi, and is author of *Argentina's Nuclear Policy* and co-author of *Nuclear Deterrence and Diplomacy*.

GARRY JACOBS is Vice President of The Mothers Service Society, Pondicherry, India, Chairman of the Committee on Peace & Development of the World Academy of Art & Science and Executive Director of the International Centre for Peace and Development.

RORY MEDCALF is Programme Director, International Security, Lowy Institute for International Policy and has worked as an intelligence analyst, diplomat and journalist.

K. SUBRAHMANYAM is India's leading strategic analyst. Former Director of the Institute for Defence Studies and Analyses, New Delhi, and Convenor, National Security Advisory Board (1998-2000), he has been Chairman of several national and international committees and task forces.

Preface
A Step Towards the Faraway Land

IT WAS early in January 2008 that Air Cmde Jasjit Singh, Director, Centre for Strategic and International Studies (CSIS), asked me if we were up to organising an international conference on nuclear disarmament, an event to mark the 20th anniversary of the Action Plan for a Non-Nuclear, Non-Violent World presented by then Prime Minister Rajiv Gandhi at the Third Special Session on Disarmament of the UN General Assembly on June 9, 1988. Rather enthusiastically and little realising the amount of work it would entail for a small organisation like the CSIS, I readily expressed my support for the idea. Fortunately, the Ministry of External Affairs stepped in to sponsor the conference and the Indian Council of World Affairs (ICWA) shared the myriad tasks as we went along.

The journey, initially, seemed all uphill – akin to what it will be in case of moving towards a world free of nuclear weapons. The Director was taken seriously unwell in February 2008 and had to undergo emergency surgery. Through March and April, the state of his health caused much concern, even as the Additional Director, AVM Kapil Kak valiantly held fort. It was on April 1, 2008, ironically April Fools Day, that we at the CSIS sent out the first invitations to the speakers and participants. Responses began to pour in within days and we were on our way, actually overwhelmed by the interest the subject was evoking around the world.

Through frenzied activity and many a challenge familiar to those who have organised international conferences, the motley team of no more than six people at the CSIS and about the same at the ICWA, worked for exactly ten weeks to D-day. Finally, June 9-10, 2008, came and went, and all the participants today remember it as a "conference of substance".

What did the conference achieve? It explored the opportunities for, and challenges of, a nuclear weapon free world (NWFW) in a transformed security context of the new millennium. It may be recalled that it was at the end of the bipolar confrontation in the early 1990s that euphoria had built around the possibility of universal elimination of nuclear weapons. However, ever since that period of optimism died down some time in the second half of that decade itself, the international community has paid little serious attention to the subject. Rather, contemporary nuclear weapons thinking and doctrines of all nuclear weapon possessors indicate an ever greater salience of nuclear weapons in their national security strategies.

Therefore, it was rewarding when experts from all over the world put their heads together to find the right point and motivation to start the journey towards a state of zero nuclear weapons. The 11 scholarly papers presented at the conference and as reproduced in this volume form an informed mosaic of ideas. They capture the richness of the debate – in range, scope, and depth. On a subject as knotty as nuclear disarmament, it is very important that citizens from all parts of the world, nuclear and non-nuclear, remain in constant engagement, exchange views and explore new ideas. Those of us who are convinced that an NWFW is indeed desirable must keep pushing for it to become feasible at some point for our future generations.

Our gratitude to the Ministry of External Affairs, Government of India, for facilitating the creation of such a platform through their generous material and moral support. Without their help and assistance, and the personal interest of the Prime Minister, Dr Manmohan Singh, the conference could not have managed the scale and level of participation that it finally did. A special word of thanks to Mr Ashok Kumar, Acting Director General, ICWA who put the weight of his organisation behind the conference.

As was evident in the two-day deliberations, the journey towards a world free of nuclear weapons, in its initial leg, would entail passing through dense forests of mistrust, dark tunnels of cynicism and frosty waters of frustration borne out of inaction. But, we certainly will find it easier to move ahead if only we could shed some of the baggage of mindsets of the past and decide to travel light. In fact, one might be surprised at the momentum that could quickly build up to reach a critical tipping point towards such a state if the political will from the top could coalesce with the push of public opinion from below.

We shall continue trying to move further down this path and to accumulate more footfalls through constant dialogue and action for the creation of a climate that will enable mankind to break free of the shackles of a weapon that has begun to control its destiny rather than the other way around. ■

Manpreet Sethi
Senior Fellow
Centre for Strategic and International Studies
New Delhi

The Third Nuclear Wave: Introductory Remarks to the New Delhi Conference

☐ **JASJIT SINGH**

HON'BLE PRIME MINISTER Dr. Manmohan Singh, Mr. Sergio de Queiroz Duarte, High Representative for Disarmament Affairs, United Nations, distinguished delegates, especially our esteemed experts from abroad, my colleague Mr. Ashok Kumar, Additional Secretary and Acting Director General of Indian Council of World Affairs (ICWA), friends, and colleagues from the Centre for Strategic and International Studies (CSIS). On behalf of ICWA and CSIS may I welcome all Cabinet Ministers, serving and former Chiefs of Air Staff, Secretaries, distinguished speakers, and participants of today and tomorrow's conference.

As you would recall, on this day 20 years ago, the then Prime Minister of India, Shri Rajiv Gandhi, presented a comprehensive Action Plan to the United Nations, not for just nuclear disarmament but also for establishing a non-violent world free of nuclear weapons. The timing was crucial. The Cold War and its numerous bush fires across the world from Central America to Afghanistan were winding down, and decades of hostilities were being replaced with new dialogue and experience sharing. In the words of Charles Dickens, it was the best of times and the worst of times: the best because we were looking at the emerging historic opportunities to reshape the world into a better and peaceful one; and yet the worst of times because old hostilities and mindsets still hovered and new challenges threatened hope.

As we know from history, tectonic changes started to impact the international landscape. And in the turmoil of the times, the world forgot about thinking and working toward a non-nuclear and non-violent world. It did not receive the type of attention it deserved – and should have received. But perhaps that was understandable. The palpable dangers of the heightened Cold War and the prospects of a catastrophic nuclear weapons exchange suddenly became remote as relations between the two superpowers began to change. The Soviet collapse soon after, in fact, altered the dynamics even further. The robust pursuit of nuclear disarmament by civil society across the world suddenly began to relax. The *Bulletin of Atomic Scientists* moved back its doomsday clock somewhat. Apparently no longer under the shadow of the bomb, the international community began to live in complacency.

Twenty years later, we find the world has changed markedly; except that nuclear weapons remain a persistent threat to mankind and its civilisation. On the other hand, nuclear proliferation took on a new dimension, spreading from a potentially unstable country in our neighbourhood to countries and regimes as far apart as North Korea and Libya. The size of nuclear arsenals started to come down; but it has been cold comfort to know that the world can be destroyed by existing arsenals only twenty times over in one stroke compared to the sixty times during the Cold War! And nuclear weapon systems and strategies began to be refined and modernised with ever greater attention to make them usable. Ballistic missile defences add to that process, promising to make nuclear war-fighting a distinct probability without the assurance of their defensive potential, but certain to alter the fragile nuclear deterrence paradigm that has existed for decades, nullifying the 1986 Reagan-Gorbachev dictum at Reykjavik. In turn, the arms race, with increasing prospects of weaponisation of space is assuming serious dimensions. Nuclear terrorism is now a distinct danger, while terrorism under a nuclear umbrella has already killed more than a hundred thousand innocents.

What this implies, ladies and gentlemen, is that we are truly into a third nuclear wave in a polarising world.

It is in this context that we felt that there is a need to refocus not only on the Action Plan of that period but also now to view it in the context of today and the future. We believe that the international community has been far too complacent about what is actually humanity's most serious threat ever since man stood upright. Hence, the effort to try and see what contribution we can make as thinkers and concerned citizens of the world since nuclear weapons are still around and new dangers have surfaced that were not even thought of that time.

Mr. Prime Minister, in this room we have assembled some of the best thinkers, policy-makers and strategic experts of the world and India on the subject and its related issues. We plan to explore the nuclear environment of the future in its diverse dimensions and discuss ways and means of dealing with them. We hope that this process will begin to lead to at least making life, if not safer for us, then at least for the coming generations. As we know, consciousness of the danger is a prerequisite to our ability to deal with it successfully. It is the consciousness slipping back in our minds that led to lowered attention to abolition of nuclear weapons after the end of the Cold War.

I am reminded of the ancient Indian political thought which laid down the duties and responsibility of *dharma* – the duty, of the *raja* – the king, the ruler. His main responsibility was to ensure the prosperity and happiness of his people; this was the social contract that existed amongst the ruler and the ruled in our country 3,000 years ago. Existence of nuclear weapons and concepts of their potential use could hardly be claimed to be consistent with such social contract which has been spreading across the world. As long as nuclear weapons exist, the risk of their use by accident, miscalculation and/or intent can never be ruled out. As long as nuclear weapons exist with some countries, other countries would find a powerful incentive to acquire them. It is in this context that we find that a nuclear weapon free world, or nuclear disarmament, as it is normally referred to, is not only linked up with the proliferation issue but is actually extremely complex. If we are serious about controlling the genie and abolishing such weapons, we will need the political resolve among the nuclear weapon states and those that could become one. That political resolve requires rational, objective, expert and intellectual support. At the very least, this needs to spell out, without bias and parochial beliefs, the main challenges of the future related to nuclear violence.

Nuclear disarmament is a complex issue; and complexity actually requires a multiple set of approaches to issues including many that apparently do not deal with nuclear weapons but with a large amount of other things. For example, one has to devote serious attention to security without nuclear weapons. And the process and the product have to be commonly agreed to. How do we get to understand the common yardstick of what we are looking for? Is it only 25,000 weapons down to 10,000 weapons and then to 1,000 weapons? Or is the process to lead to a global zero? But when we look closely, it is also something much more than just the numbers. What is crucial is the attitude towards weapons of mass destruction that requires our attention. We need to remember that major changes to deeply entrenched attitudes over long periods have changed only after the belief systems began to change. Abolition of slavery, untouchability, apartheid and even colonialism started to become realities not on the basis of treaties and laws, but through change in the belief systems. This offers us a starting point in the right direction.

Secondly, nuclear deterrence has been justified, legalised and legitimised in every different way for the past six decades. Because it has not broken down, nuclear deterrence cannot be assumed to be the bedrock of future peace and security. Deterrence, of course, has always been part of human history and life. But the dynamics of nuclear deterrence have changed in the past two decades. Who to deter, and how to deter some of the adversaries now pose serious questions. The issue is not about rogue states, but of more complex situations.

For example, cultures and nations that promote and venerate the ideology and practice of suicide bombing challenge the traditional concept of deterrence.

Similarly, we observe that one of the major developments in the past two decades is trans-national terrorism, made more vicious by religious radicalism, and insecurity and vulnerability of even the sole superpower caused by hijacked unarmed civil aircraft with innocent passengers inside leading to what can only be called horrendous mass destruction.

You will now recall the last century, Mr. Prime Minister, had been the most violent century in human history. Surely, we do not want the 21st century like that. We must make a contribution; we are not going to change the fundamental thinking of a lot of people. That remains a difficulty and a problem. But we must make the effort. We must make the effort not only in a larger perspective that the world should be safer than it is because now what is required is much more attention to human development, to enhancing the quality of life and happiness.

Yet self-defence remains a fundamental right of human beings, as well as every nation.

Ladies and gentlemen, as you know, controlling nuclear proliferation is an intrinsic aspect of any measure toward a nuclear weapon free world. But proliferation itself has changed in many ways in recent decades. Non-proliferation measures and regimes apparently have long ago reached their plateau. The international community appears to be at a loss to find new approaches. The World Court had ruled in 1996 that no law beyond the general law governing armed conflict applies to nuclear weapons. Possession of nuclear weapons by some countries provides an incentive to others that seek self-defence against such weapons. And we are now told that a single scientist built up a nuclear bazaar to sell, barter and transfer nuclear weapons materials and designs to other countries for years in spite of the army in his country claiming full control over its nuclear assets; and very little was or could be done about it.

As we move toward more micro nukes without giving up the big bombs, nuclear weapons become much more usable because more options would be available to apply destructive diplomacy; and we need to look at that very carefully. The doctrines and strategy of first-use, ready to fire, launch on warning, launch under attack, and all the doctrinal sophistry imply that the time for sober reflection before giving the order to launch nuclear weapons has contracted enormously. Of course, China and India have provided a new model; but this has not been followed by the dominant powers in spite of the fact that their security calculus has changed dramatically from the Cold War period. We need to examine this model rather than reject it out of hand as has been done, from the perspective of whether a counter-strike doctrine and strategy would not serve the legitimate self-defence needs of a nation threatened by nuclear weapons, as an interim model for the future till we get to a nuclear weapon free world.

Another major development under way has been the revival and resurgence of nuclear power. There are many reasons for it, ranging from the need for assured environment friendly energy production to the dwindling hydrocarbon reserves, especially in the context of rapidly rising consumption so critical to human development in developing countries. But the technology and materials for peaceful uses and for weapons are not far apart. The spread of nuclear power and its technology would naturally add to proliferation concerns. As long as the pressure to acquire nuclear power reactors had remained limited and the developing countries had few options in the face of restrictive technology access regimes, the risk was seen as manageable. But with the rapid increase in demand and plans for robust expansion of nuclear energy across the world, the implications for proliferation have become more complex. We have to seek effective ways to reconcile nuclear power with the lowered risks of proliferation.

But we also recognise that political resolve – and not just the legal, technical processes – would finally shape the future of nuclear weapons. Hopefully, the phase of complacency after the end of Cold War is truly over. The world is on the verge of moving toward polarisation. New threats to the security of states and their citizens abound. Targeting and killing of innocents appear to have been legitimised by a number of countries. There is an identifiable shift in global power from the West to the East. Globalisation creates its own challenges and assets. There are challenges like global warming and climate change that simply cannot be addressed only at national levels. An international cooperative approach has become an imperative; and a global security framework has to be recast in a new paradigm to meet the challenges of the 21st century. Nuclear weapons pose one of them, albeit the most destructive one.

Two issues deserve our attention here. One is that ever since the nation state began to emerge in the international system after the Treaty of Westphalia in the 17th century, international security began to be founded on the foundations of competition between states. Later developments affecting key functions of the state, like market economy, electoral democracy, military power and technological advancement all have been competitive enterprises. The principle of sovereignty demands that a state maximise its security and power in relation to other states, many of which would face increased insecurity and are compelled to enhance their security, even if with nuclear weapons, leading to an upward spiral of competition. But current and future challenges in a globalising world cannot be managed effectively without international cooperation. Hence, a paradigm shift from competitive security to cooperative security is critical for managing future challenges.

Another aspect is that of our goals. Prime Minister Nehru used to emphasise the goal of peace in preference to that of security. An environment of peace would naturally provide security, whereas mere security may or may not bring peace. For example, security in Europe during the Cold War was ensured for 45 years by something like 60,000 nuclear weapons, 94,000 combat aeroplanes, about 110,000 tanks and massive quantities of other weapons and military systems. But only now, for the last 15-16 years, they are searching for peace. Peace has to be given a chance in shaping future paradigms. It is not just a question of semantics. It is a question of asking: what is the objective that we are seeking in terms of a long-term perspective?

What we need is a fundamental conceptual change in terms of a fundamental paradigm of inter-state security and national security. Therefore, how do we manage competitive security now in a world which I am afraid, must be competitive in different ways, but which must cooperate in an increasing number of areas of vital concern to humanity and its future? In that cooperative security paradigm, what should be the role of nuclear weapons? Obviously, it must be the very minimum, lowest level of restraint that the human mind is capable of ensuring, and vigorous efforts to abolish nuclear weapons.

It is from this perspective, Mr. Prime Minister, that we have organised this conference and we hope to take on specific issues for professional examination. The difference with the current nuclear phase compared to the earlier attempts is that policy-makers and strategic experts across the world (rather than peace movements, etc) have started taking new interest partly because of some of the things that I have mentioned and partly for other equally compelling reasons. Our agenda here is to examine, discuss, debate and try to find ways that would work for the future to control the growing menace of nuclear weapons and abolish them at an early date while paying serious attention to managing international peace and security in a non-nuclear world.

We hope Mr. Prime Minister that you will give us the guidance to take off from this conference not only for the deliberations for the two days but beyond that, for India to take on its natural responsibility in a world which should be hopefully more peaceful and free of nuclear weapons.

May I invite you Mr. Prime Minister to deliver the Inaugural Address? ∎

Inaugural Address

☐ **MANMOHAN SINGH**

WELCOME to New Delhi. We take great pride in hosting this international conference on a subject that touches upon the very survival of humankind.

Twenty years ago, on this day, our former Prime Minister, Shri Rajiv Gandhi addressed the Third Special Session on Disarmament of the UN General Assembly. Speaking on the theme of "A World Free of Nuclear Weapons," he introduced an Action Plan calling on the international community to negotiate a binding agreement on general and complete disarmament. At the heart of the Action Plan was a commitment to eliminate all nuclear weapons in three stages by 2010.

Rajiv Gandhi believed that disarmament, in particular nuclear disarmament, was essential to usher in a safe and non-violent world. He had a deep insight into the nature of evolution of technology, its potential for advancing human welfare as also for unleashing destruction. In this context, he was acutely aware of the power of the atom. He wished that it should never again be used for destructive purposes.

The Rajiv Gandhi Action Plan was a comprehensive exposition of India's approach towards global disarmament and continuity in our thinking. It symbolised the continuity in our thinking since 1954, when India pioneered the call for a complete ban on nuclear testing. The essential features of the Action Plan continue to remain valid even today.

A review of developments since 1988 presents a mixed picture on how far the world has moved to realise the vision of Rajiv Gandhi. On the one hand, the end of the Cold War has created an opportunity for the world to move away from the dangerous doctrines that were based on the precept of Mutually Assured Destruction. Greater engagement and realisation of the inter-dependent nature of global security among major powers has created new space for action on disarmament. In 1993, the Conference on Disarmament was able to finalise the Chemical Weapons Convention, a multilateral, non-discriminatory and internationally verifiable treaty aimed at eliminating an entire category of weapons of mass destruction by a fixed date, namely 2012.

On the other hand, the painful reality is that the goal of global disarmament, based on the principles of universality, non-discrimination and effective compliance, still remains a distant one.

Even more disturbing, however, is the emergence of new threats and challenges to global security. I refer to the growing risk that nuclear weapons may be acquired by terrorists or those driven by extreme ideologies; the increasing danger of non-state actors accessing nuclear materials and devices; the development of new weapon systems based on emerging technologies which pose challenges to space security and provide new roles for nuclear weapons; and the weakening of multilateralism even as bilateral arms control processes falter in shifting strategic landscapes.

The threat of climate change and global warming itself raises a range of security concerns, especially for us in the developing world.

India, which has witnessed rapid economic growth in the last few years, and is poised for even higher growth rates in the future, needs a peaceful international environment so that we can focus our resources on improving the lives of our people. We seek a world in which power flows through the empowerment of people; and from the strength and resilience of our economy, our society, our institutions and our values.

Our energy needs will continue to rise in the foreseeable future. We do not have the luxury of limiting our options of energy sources. We, therefore, wish to create an international environment in which nuclear technology is used not for destructive purposes but for helping us meet our national development goals and our energy security.

India is fully aware of its responsibilities as a nuclear weapon state. We have a declared doctrine of no first use that is based on credible minimum deterrence. We have in place strict controls on export of nuclear and fissile related materials and technology. India has no intention to engage in an arms race with anyone. Above all, India is fully committed to nuclear disarmament that is global, universal and non-discriminatory in nature. The pursuit of this goal will enhance not only our security but the security of all other countries.

These objectives cannot be achieved through partial methods and approaches. The only effective form of nuclear disarmament and elimination of nuclear weapons is global disarmament. Nuclear weapons know no boundaries. Even today, the nuclear arsenals in possession of the major powers are enough to destroy the world many times over. In this scenario, it is not possible to "regionalise" nuclear disarmament.

It is in keeping with this approach that India has recently submitted a Working Paper on Nuclear Disarmament to the UN General Assembly, containing initiatives on nuclear disarmament. We hope to stimulate a debate and promote consensus on the way forward. These proposals have also been submitted before the Conference on Disarmament in Geneva. They are a set of practical measures for working towards the goal of a nuclear weapon free world. We do not wish to exclude other measures that may contribute to achieving this goal, nor do we hold that there is a rigid hierarchy among these steps and a specific sequencing for their implementation. The measures we suggest include:

- Reaffirmation of the unequivocal commitment of all nuclear weapon states to the goal of complete elimination of nuclear weapons.
- Reduction of the salience of nuclear weapons in security doctrines;
- Adoption of measures by nuclear weapon states to reduce nuclear danger, including the risks of accidental use of nuclear weapons;
- Negotiation of a global agreement among nuclear weapon states on "no first use" of nuclear weapons;
- Negotiation of a universal and legally-binding agreement on non-use of nuclear weapons against non-nuclear weapon states.
- Negotiation of a convention on the complete prohibition of the use or threat of use of nuclear weapons; and
- Negotiation of a nuclear weapons convention prohibiting the development, production, stockpiling and use of nuclear weapons, and on their destruction, leading to the global, non-discriminatory and verifiable elimination of nuclear weapons with a specified time-frame.

These proposals retain the spirit and substance of the Rajiv Gandhi Action Plan. We hope that other states will agree to a dialogue on these proposals, and will join us in committing to nuclear disarmament. That is the critical first step – a commitment, preferably a binding legal commitment, through an international instrument, to eliminate nuclear weapons within a time-bound framework. In parallel with this general commitment to nuclear disarmament, we need strengthened non-proliferation commitments such as on denying nuclear material, technology and equipment to terrorists. Pending global nuclear disarmament, all states must ensure that they do not allow proliferation of sensitive technologies into dangerous hands.

India is ready to add its own weight and voice to the global debate on nuclear disarmament with a view to crafting such a consensus on disarmament and non-proliferation. We need a collective approach anchored in a universal partnership that is supported by non-governmental communities and public opinion.

I wish your deliberations all success. ■

Keynote Address

☐ **SERGIO DUARTE**

I WISH to begin by thanking the Centre for Strategic and International Studies and the Indian Council of World Affairs for organising this event and for inviting me to participate. I share the view that the Action Plan introduced by India in 1988 deserves to be remembered, both for its wisdom, and for the abiding relevance of its powerful vision of a path toward a safer and more peaceful world. It is not surprising that such a vision would emerge from a land the world has long associated with the non-violent ideals of the Buddha and Mahatma Gandhi.

In my invitation, the organisers said that the aim of this conference is to "spur dialogue on nuclear disarmament as a strategic necessity of our times." After witnessing many years of setbacks in this field, I have come to the conclusion that the world has, at long last, reached a point where such a dialogue can finally take place productively and in good faith. Any optimism in this regard must, of course, be tempered by a sober recognition of the work that lies ahead to overcome many persisting obstacles.

In introducing his Action Plan to the UN General Assembly on June 9, 1988, Prime Minister Rajiv Gandhi had a lot to say about nuclear deterrence. He called it the "ultimate expression of the philosophy of terrorism,"a doctrine that, in his words, "will take us like lemmings to our own suicide." Over the years, he was, of course, not the only one to have made this connection between nuclear deterrence and terrorism. The 2006 report of the Weapons of Mass Destruction Commission, chaired by Hans Blix, bears the title, "Weapons of Terror."

Yet today, as we mark the 20th anniversary of the Action Plan, we have not five but about eight or nine states that possess nuclear weapons and that adhere to some version of the doctrine of nuclear deterrence. While this number is still small – surely relative to the over 180 states that have chosen not to acquire such weapons – it is troubling indeed that over half of the world's population now lives in states that are armed with nuclear weapons or are covered by nuclear umbrellas.

This year also brings the 40th anniversary of the signing of the Treaty on the Non-Proliferation of Nuclear Weapons (NPT). Its obligations included a legal commitment "to pursue negotiations in good faith on effective measures relating to cessation of the nuclear arms race at an early date and to nuclear disarmament, and on a Treaty on general and complete disarmament under strict and effective international control." Though nobody knows exactly how many nuclear weapons exist, the number is often estimated around 26,000. We are surely well past the NPT's "early date" for negotiations on nuclear disarmament and the world is still struggling to confront three nuclear threats – those associated with existing arsenals, the prospect of the proliferation of such weapons to additional states, and the danger of their acquisition by terrorists.

Can the Action Plan – which was reportedly soon followed by India's decision to acquire such weapons – still inspire constructive multilateral initiatives for global nuclear disarmament? Yes, I believe it can, especially if viewed in its historical context.

The world knows that India has long advocated the elimination of nuclear weapons and still does so today, along with other disarmament goals. I am very pleased that the UN's Regional Centre for Peace and Disarmament in Asia and the Pacific will open next month its office in Kathmandu, an event that has long been promoted by India – and the Centre will both need and welcome its support in the years ahead.

India's efforts for nuclear disarmament have, in fact, spanned several decades. Mahatma Gandhi himself condemned the atomic bombings in Japan, declaring that "the only weapon that can save the world is non-violence." On April 2, 1954, Prime Minister Nehru responded to the tests of hydrogen bombs by calling upon the United States and the Soviet Union to conclude a "standstill agreement" on further tests pending progress in disarmament. India presented this proposal to Secretary-General Dag Hammarskjöld a few days later and has ever since been a persistent advocate in the UN of nuclear disarmament, the peaceful uses of nuclear energy, and a comprehensive nuclear test ban treaty, which I hope will soon enter into force.

India's interest in "general and complete disarmament" is also important to note in this context. Due to mandates derived from the Charter and early General Assembly resolutions, the UN has simultaneously pursued the elimination of all weapons of mass destruction and the regulation of conventional armaments. In 1959, the General Assembly adopted a Soviet resolution that combined these into the concept of "general and complete disarmament under effective international control." The Soviet plan offered a three-stage approach to disarmament, to be implemented in four years – the United Kingdom had also issued a three-stage approach to disarmament in September of that year. In 1961, the United States and the Soviet Union announced agreement on a Joint Statement of Agreed Principles for a treaty on general and complete disarmament that would be achieved "in an agreed sequence, by stages."

Later that year, the General Assembly created the Eighteen Nation Committee on Disarmament (or ENDC), a negotiating body that included India as a new member, and that proceeded to consider treaty proposals offered by the United States and Soviet Union. Each of these proposals included a three-stage approach to achieving general and complete disarmament. In those deliberations, India proposed a draft preamble to such a treaty, which included several goals that were closely aligned with the Charter, including the peaceful resolution of disputes and an agreement to renounce war as an instrument of policy. I am noting these developments because both the three-stage approach and the renunciation of war later figured prominently in the 1988 Action Plan.

Throughout the decade of the 1960s, India contributed actively to deliberations in the ENDC. I personally witnessed these contributions as a junior member of the Brazilian delegation to the committee. I recall how the negotiating priorities shifted in the mid-1960s from an emphasis on pursuing a treaty on general and complete disarmament, to what were then called "partial measures," which included the non-proliferation of nuclear weapons and a comprehensive ban on nuclear tests.

In this period, India joined what was known as the "Group of Eight" – a small but diverse group of states on the committee sharing common views on disarmament issues. This group issued a series of Joint Memoranda describing such views in some detail. One such memorandum, dated September 15, 1965, stated that a "treaty on non-proliferation of nuclear weapons is not an end in itself but only a means to an end" namely, the achievement of nuclear disarmament. Throughout these years, India's representative in the ENDC argued that a true non-proliferation treaty must cover both the improvement or expansion of existing arsenals (that is, vertical proliferation) and the spread of such weapons to additional states. That view was actively supported by the other members of the Group of Eight.

On November 19, 1965, the General Assembly adopted Resolution 2028, which called on the ENDC to negotiate a nuclear non-proliferation treaty that embodied five principles: it should be void of any loopholes; it should "embody an acceptable balance of mutual responsibilities and obligations"; it should be a step toward general and complete disarmament; it should contain provisions to ensure the effectiveness of the treaty; and it should not interfere with the right of states to pursue regional nuclear weapon-free zones.

India later objected that the text of the NPT did not satisfy these criteria because of the different obligations of the nuclear weapon and non-nuclear weapon states. India argued that the former should stop manufacturing such weapons, as the latter commit not to acquire them. India also claimed that the right to peaceful uses of nuclear energy should extend to the right of non-nuclear weapon states to conduct nuclear explosions for peaceful purposes. At the ENDC in 1966, India's ambassador declared, "The so-called balance of terror seems to produce only terror but no balance, and provides proliferation but no stable security."

On May 23, 1967, India elaborated its views in the ENDC on the concept of a non-proliferation treaty, saying that such a treaty "can be a purposeful instrument only if this negotiating Committee of ours conceives of that measure in the overall and universal concept of disarmament and not as a simple exercise in imposed non-armament of unarmed countries." The same statement regretted that "no real or effective effort is being made to deny prestige to possession of nuclear weapons." Despite such concerns, the NPT was later opened for signature on July 1, 1968. Though I have recalled today several Indian statements, I must add that similar concerns were voiced by the other members of the Group of Eight.

By the late 1970s, especially after India detonated its "peaceful nuclear device" in 1974, growing international concerns about the threat of nuclear weapons proliferation, expanding nuclear arsenals, and a deterioration of relations between the two superpowers, inspired new approaches – both from governments and civil society – to revitalise disarmament efforts. These concerns culminated in the General Assembly's first special session on disarmament in 1978, which adopted a detailed international disarmament strategy covering both goals and the institutions needed to achieve them – which are together known as the UN disarmament machinery.

Global demands for progress in disarmament were clearly growing in those years. In 1980, Sweden established the Independent Commission on Disarmament and Security Issues under former Prime Minister Olof Palme, with India's Ambassador C.B. Muthamma as one of its members. Its final report was issued on June 1, 1982, which identified several short-term and medium-term measures to advance general and complete disarmament, covering weapons of mass destruction, conventional arms, and various measures to strengthen the UN and to advance the common theme of "common security." Two weeks later, an estimated one million people participated in a march in New York City for global nuclear disarmament.

In May 1984, Palme joined with several other heads of state and government – including India's Prime Minister Indira Gandhi – to launch the "Six-Nation Initiative," an idea promoted by an international network then called "Parliamentarians for World Order." I am pleased to recognise Senator Douglas Roche of Canada – the first international president of this group who is attending our conference today. This coalition of middle-power countries began many years of efforts to promote progress in nuclear disarmament, achieve a comprehensive nuclear test ban, and advance other goals relating to general and complete disarmament. Its efforts focussed on promoting détente among the nuclear-weapon states – recognising that the responsibility to avert a global nuclear catastrophe did not rest with such states alone.

Rajiv Gandhi succeeded his mother as a member of the Six-Nation Initiative following her assassination in October 1984; he also served as a member of the Palme Commission, which continued to monitor disarmament issues in the 1980s after publication of its report.

The Six-Nation leaders issued several joint declarations in this period to advance their goals. The Delhi Declaration of January 28, 1985, focussed on the prevention of an arms race in outer space and the need for a nuclear test ban. It concluded, "Progress in disarmament can only be achieved with an informed public applying strong pressure on governments."

On January 19, 1986, the Palme Commission met in India and issued a Delhi Statement, which reiterated the need for action on many of the goals of its 1982 report, including nuclear disarmament and non-proliferation, the nuclear test ban, outlawing space weapons, limiting conventional arms, strengthening the UN, and promoting regional security.

Shortly before his assassination, Olof Palme and the other Six-Nation leaders issued a joint message of February 28, 1986, to Presidents Reagan and Gorbachev, calling for a moratorium on nuclear tests. A nuclear test ban was also stressed in the Mexico Declaration of August 7, 1986, by the Six-Nation leaders. On November 27, 1986, Rajiv Gandhi and President Gorbachev signed a Joint Declaration of Principles of a Nuclear Weapon Free and Non-Violent World, echoing many of the themes relating to general and complete disarmament, including a proposal for the elimination of nuclear weapons "before the end of the century" and for the progress toward a "nuclear weapon free civilisation."

Therefore, by the time Rajiv Gandhi launched his Action Plan in the General Assembly, many of its components had already gained widespread recognition and support throughout the world. While the plan emphasised global nuclear disarmament, it also had several broader objectives that together, in Rajiv's words, constituted "ideas for the conduct of international relations in a world free of nuclear weapons."

The plan proposed the elimination of nuclear weapons in a three-stage process that was to conclude no later than the year 2010, by means of a nuclear-weapons treaty to replace the NPT. Under the plan, states would have to make tangible progress in achieving the goals of one stage before proceeding to the next. All states must be part of the disarmament process, which would also aim at establishing a broader "Comprehensive Global Security System" involving doctrines, policies, and institutions to sustain a world free of nuclear weapons. The plan identified goals for deep reductions in nuclear arsenals, a cut-off in the production of fissile material for weapons, a test ban, a prohibition on the use or threat of use of nuclear weapons, and for the establishment of "an integrated multilateral verification system under the aegis of the United Nations."

The plan did not stop with nuclear weapons and their delivery systems. It called for treaties banning chemical and radiological weapons, limitations in conventional forces, a moratorium and ban on space weapons, measures to prevent the development of weapons based on new technologies, and a shift of resources out of armaments into economic development.

In light of the history I have covered, it is clear that this Action Plan is very much consistent with over six decades of work both inside and outside the United Nations to advance the internationally agreed goal of general and complete disarmament. It most notably applies to the aim of eliminating all weapons of mass destruction and the limitation of conventional armaments. Yet it goes somewhat further by calling for various measures to strengthen the parts of the Charter dealing with the peaceful resolution of disputes and the avoidance of threats of the use of force. It offers some specific ideas on institutional development, notably the proposal for a role for the UN in the field of verification.

It is, of course, possible to update this Action Plan – I note, for example, that some of its measures have already been achieved, like the Chemical Weapon Convention and the conclusion of negotiations on a CTBT. Yet what I find most striking is how many of the initiatives in this plan remain relevant today. In fact, if the Conference on Disarmament (CD) in Geneva were to finally be able to proceed to negotiate a treaty on general and complete disarmament, it would almost certainly have to cover many if not most of these initiatives. How could such a treaty, for example, neglect the disposition of fissile material, delivery systems, or the sequencing of deep cuts in nuclear arsenals?

Let us recall that both the General Assembly and the states parties to the NPT have repeatedly identified specific criteria that must apply to future efforts in the field of nuclear disarmament – criteria including verification, transparency, irreversibility, and the need to move forward through binding legal commitments. It seems to me that these are precisely the type of criteria that were proposed in the 1988 Action Plan. I also find that the plan is fully consistent with the common sense approach adopted for decades in the pursuit of general and complete disarmament, namely, that both nuclear disarmament and conventional arms limitations should be pursued simultaneously, just as the goals of disarmament and non-proliferation must also be advanced together rather than sequentially.

The true goal of general and complete disarmament is not disarmament itself, but the enhanced security of all states. Here is how the General Assembly framed this goal at its first special session on disarmament in 1978, in language that remains widely accepted today:

> Together with negotiations on nuclear disarmament measures, negotiations should be carried out on the balanced reduction of armed forces and of conventional armaments, based on the principle of undiminished security of the parties with a view to promoting or enhancing stability at a lower military level, taking into account the need of all States to protect their security.

This language corresponds well with the basic approach articulated in the 1988 Action Plan.

Our work today should, however, not be limited to the act of retrospection. The 1988 Action Plan identified several worthy goals for the world community to pursue. It did not, however, provide much guidance on the specific means that would be needed to achieve them. What, for example, is needed to raise the priority of disarmament in key states and to break the stalemate at the CD in Geneva? The answers will, of course, vary with the different political and technological circumstances among the various states. Despite this diversity, I believe the course ahead will depend heavily upon initiatives from civil society and concerned governments throughout the world.

As Rajiv Gandhi stated in his speech to the General Assembly in 1988, "The battle for peace, disarmament and development must be waged both within this Assembly and outside by the peoples of the world ... [adding that] The ultimate power to bring about change rests with the people." This conclusion certainly points us in the right direction. Let us recall that a network of parliamentarians led to the Six-Nation Initiative. Today, we have the seven-nation New Agenda Coalition and the seven-nation Norwegian initiative, which are both working to advance many of the goals also found in the Action Plan.

These initiatives are in turn supported by non-governmental organisations (NGOs) around the world. In Japan, the Mayors for Peace initiative has gained the support of 2,277 mayors in 129 countries, to advance their own "Overall Action Plan" for peace and disarmament. In the United States, we have numerous NGOs working to advance nuclear disarmament goals, in a collective effort that has been revitalised by the publication of opinion-editorials by former US officials George Shultz, Henry Kissinger, William Perry, and Sam Nunn advancing the "Hoover Plan" for global nuclear disarmament.

These are just some of the various plans and initiatives that are very much alive in the pursuit of global nuclear disarmament. They deserve our recognition and our support.

I cannot speculate on the extent that these more recent plans have been inspired by ideas from the 1988 Action Plan, but I am sure that progress in this field is cumulative. Ideas from civil society or within governments start circulating in General Assembly debates and appearing in resolutions, and are actively pursed by groups of states, networks of non-governmental groups and individuals, promoted in interesting ways, elaborated and adapted to respond to changing conditions, and advanced through a shared conviction that the cause is right and in the common interest.

As an approach to international peace and security, disarmament has always embodied a fusion of ideals and self-interest. It is not just a dream, but a practical means to achieve security – one that is far more reliable than its alternatives, including nuclear deterrence, the balance of power, or self-help. Much has been written about how nuclear weapons have been pursued for reasons of security and prestige. I believe we are now entering an age in which nations begin to realise that the same goals are served not by the acquisition, but by the elimination of such weapons, and served far better. Nuclear disarmament unites what is right with what works. This is why it has become, in the language of this conference, a "strategic necessity."

As a former pilot, perhaps Rajiv Gandhi had flown over too many countries without witnessing any walls or barbed-wire fences. Perhaps he saw the globe as a whole, a place where people of all nationalities share some common values, common purposes, and common interests. Though most of us are not pilots or astronauts, we too can understand the extent that the interests of individual states are a function of the collective interest and welfare of the community of states, or more precisely, the peoples of the world. I believe this was the basic concept that inspired the 1988 Action Plan. It is an idea well worth the name of action now, twenty years later, in 2008.

In closing, I would like to recall the words of Rikhi Jaipal, who served as the secretary – and the secretary-general's personal representative – to the Committee on Disarmament in Geneva in the early 1980s, during my second tour of duty there with the Brazilian mission. In his 1986 book, *Nuclear Arms and the Human Race*, he warned that:

... the danger of nuclear war will remain with us as long as there are people who think that the nuclear weapon cannot be disinvented. They should be aware that the human race cannot be reinvented after a nuclear holocaust.

His words are an eloquent reminder that the key to future progress in disarmament remains in the realm of reshaping human perceptions – including perceptions of the inherent dangers in possessing such weapons, perceptions of their risk of use, and perceptions of the concrete security benefits that would flow from their global elimination. These perceptions will help us to navigate the road leading to a world free of nuclear weapons. Let us today recommit ourselves to achieving this great and historic goal. ■

Special Address

☐ **PRANAB MUKHERJEE**

I WELCOME you all this evening. I am particularly thankful to all our guests from overseas who have joined these two-day deliberations on a subject of seminal importance.

The international community has long recognised that nuclear weapons pose the greatest danger to mankind. The UN General Assembly, in its very first resolution, set the goal of elimination from national armaments of atomic weapons and of the use of atomic energy for peaceful purposes only. This goal has been reaffirmed many times since then, notably at the First Special Session of the General Assembly on Disarmament, to which we owe the current disarmament machinery. The Non-Aligned Movement too has emphasised the necessity to start negotiations on a phased programme for the complete elimination of nuclear weapons within a specified framework of time. The International Court of Justice concluded in 1996 the there exists an obligation to pursue in good faith and bring to a conclusion negotiations leading to nuclear disarmament under strict and effective international control.

Despite these expressions of global opinion, the end of the Cold War and some moves by the largest possessors – the United States and Russia – to reduce their stockpiles, we are still a long way from complete and verifiable elimination of nuclear weapons. Today, the problem is rendered infinitely more complex by the spread of international terrorism and clandestine proliferation.

International efforts towards a nuclear weapon free world can be fruitful only when backed by a solid and sustainable consensus of all states. The forums are there, the tools are known and many of the pathways, notably that of delegitimisation, are familiar to all. What is required is a commitment. In India, we are willing and our commitment is firm. This is a commitment that has been expressed by many of our leaders, but perhaps by none more eloquently and comprehensively than the late Shri Rajiv Gandhi. The speech and the Action Plan presented by Rajiv Gandhi to the United Nations Special

Session on Disarmament on June 9,1988, remains a guiding principle of India's policy on disarmament and non-proliferation. As the prime minister remarked this morning we remain committed to the central tenet of that plan, namely, the complete elimination of nuclear weapons, leading to global disarmament in a time-bound framework.

All of you in this distinguished gathering are aware that unlike 1988, India today is a nuclear weapon state. However, this does not diminish in any manner our commitment to nuclear disarmament. We are ready to support multilateral negotiations leading to an early conclusion of a nuclear weapons convention prohibiting the development, production, testing, deployment, stockpiling, threat or use of nuclear weapons and providing for their elimination.

Meanwhile, practical steps that reduce the danger of nuclear war posed by the present hair trigger alert of nuclear weapons can be contemplated. These steps would also advance the process of delegitimisation of nuclear weapons, a process that was so successfully employed in the elimination of another category of weapons of mass destruction through the Chemical Weapons Convention. These steps must include strengthened commitments on non-proliferation, especially to prevent leakage of dangerous technologies to non-state actors. In this context, the pursuit or clandestine activities in respect of nuclear technologies is unacceptable and dangerous. All states must transparently live up to the international commitments that they have undertaken.

The Rajiv Gandhi Action Plan was a holistic framework seeking negotiations for a time-bound commitment for the complete elimination of nuclear weapons to usher in a world free of nuclear weapons and rooted in non-violence. It was comprehensive, covering issues ranging from nuclear testing and production of fissile material for nuclear weapons to a time-bound elimination of nuclear stockpiles. These issues remain relevant today. On the twentieth anniversary of the plan, it may be time to build a consensus on a comprehensive disarmament and non-proliferation framework that allows the international community to initiate concrete steps towards the goal of nuclear disarmament.

I assure you that India is ready to joint efforts towards building such a consensus. ■

Towards a Nuclear Weapon Free and Non-Violent World Order

☐ **MANI SHANKAR AIYAR**

INDIA is now a nuclear weapon state (NWS). Further, it has affirmed its intention to maintain a credible minimum nuclear deterrent.

How does this affect the Action Plan for a nuclear weapon free and non-violent world order submitted by Prime Minister Rajiv Gandhi to the Third Special Session on Disarmament of the United Nations General Assembly at New York on June 19, 1988?

Soon after the nuclear weapon test at Pokhran in May 1988, the Indian National Congress affirmed that the Rajiv Gandhi Action Plan remained the "sheet anchor" of the external dimension of the party's nuclear weapons policy. The party then undertook an exercise to update and present in treaty language, a draft convention incorporating the Rajiv Gandhi Action Plan. This draft was formally submitted to the secretary general of the United Nations by the Congress president in 2001.

In his statement to Parliament on July 29, 2005, the Prime Minister, Dr. Manmohan Singh, said:

> Our commitment to work for universal nuclear disarmament, so passionately espoused by Prime Minister Rajiv Gandhi, in the long run, will remain our core concern.

Subsequently, replying to a debate in the Rajya Sabha (Upper House) on August 17, 2006, the prime minister said:

> Our commitment towards non-discriminatory global nuclear disarmament remains unwavering, in line with the Rajiv Gandhi Action Plan. There is no dilution on this count. We do not accept proposals put forward from time to time for regional non-proliferation or regional disarmament. Pending nuclear disarmament, there is no question of India joining the NPT as a non-nuclear weapon state, or accepting full-scope safeguards as a requirement for nuclear supplies to India, now or in the future.

He further added:

Our support for global nuclear disarmament remains unwavering. Prime Minster Rajiv Gandhi had put forward an Action Plan in the 1988 UNGA Special Session on Disarmament. We remain committed to the central goal of this Action Plan, that is, complete elimination of nuclear weapons leading to global nuclear disarmament in a time-bound framework.

Soon after assuming the office of Minister of External Affairs, Shri Pranab Mukherjee, in an interview to *The Hindu*, published on November 21, 2006, said:

> We are committed to non-proliferation and disarmament. What Rajiv Gandhi said at the Special Session on Disarmament of the United Nation [June 9, 1988] is the guiding principle of our foreign policy. He told the world that we would not graduate ourselves from the threshold level – that was our position then, before 1998. We want that those who have nuclear weapons should stop proliferation – vertically, horizontally – reduce stockpiles and have a time-bound action plan [for disarmament]. And that [commitment] stands. In between, of course, we have gone for the [May 1998 nuclear] explosions. There have been developments and that cannot be erased. It has already taken place – but even in that context, we are serious and we are engaging ourselves. In this United Nations session, we are going to move a resolution to this effect [for time-bound disarmament].

It will, thus, be seen that for sixty years, India has been consistent in pleading for universal disarmament:

- This was sought for the 27 years between 1947 and 1974 when India had not undertaken any nuclear test.
- It remained so after the first series of tests at Pokhran in 1974 which established India as a threshold NWS.
- And it has remained so since India became an NWS in 1998.

MAHATMA GANDHI

The origins of this consistency in pressing for universal nuclear disarmament begin with Mahatma Gandhi's revulsion at the first use of nuclear weapons at Hiroshima and Nagasaki in August 1945:

> I did not move a muscle when I first heard that the atom bomb had wiped out Hiroshima. On the contrary, I said to myself, "Unless the world now adopts non-violence, it will spell certain suicide for mankind."

The Mahatma urged:

> The moral to be legitimately drawn from the supreme tragedy of the bomb is that it will not be destroyed by counter-bombs, even as violence cannot be destroyed by counter-violence. Mankind has to get out of violence only through non-violence. Hatred can be overcome only by love. Counter-hatred

only increases the surface as well as the depth of hatred.

Writing in his magazine *Harijan*, on July 7, 1946, about a year after the atomic bombing of Hiroshima and Nagasaki, Gandhiji said:

> It is being suggested by American friends that the atom bomb will bring in *Ahimsa* (non-violence) as nothing else can. It is meant that its destructive power will so disgust the world that it will turn it away from violence for the time being. This is very like a man glutting himself with dainties to the point of nausea and turning away from it only to return with a redoubled zeal after the effect of nausea is well over. Precisely in the same manner will the world return to violence with renewed zeal after the effect of disgust is worn out.

How prescient! For a while, though from time to time, there has been a surge of public opinion and governmental concern over nuclear weapons even in nuclear weapon states, the sad fact is that such sporadic surges of anti-nuclear weapon sentiment have generally given way to acquiescence or even assertion of the need for such weapons. This only validates the next paragraph in Gandhiji's article of July 1946.

> So far as I can say, the atomic bomb has deadened the finest feeling that has sustained mankind for ages. There used to be the so-called laws of war which made it tolerable. Now know the naked truth. War knows no law except that of might.

Returning again to this theme on November 16, 1947, a few months after India became independent, Gandhiji wrote:

> In this age of the atom bomb, unadulterated non-violence is the only force that can confound the tricks of violence put together.

When asked whether the atomic bomb had rendered non-violence obsolete, Gandhiji returned an emphatic "No", adding:

> On the contrary, non-violence is the only thing that is left in the field. It is the only thing that the atom bomb cannot destroy.

JAWAHARLAL NEHRU

Building on the legacy of the Mahatma, Jawaharlal Nehru envisaged a twin-track policy. On the one hand, India had to harness its scientific and technological talent in the campaign for what was then called "Atoms for Peace" and, on the other, India had to be in the forefront also of the campaign for nuclear disarmament.

Indeed, as early as 1940, Jawaharlal Nehru, in a confidential note penned for the use of the inner councils of the Congress Party had written:

> Both because of our adherence to the principle of non-violence and from practical considerations arising from our understanding of world events, we believe that complete disarmament of all nation states should be aimed at,

and is in fact an urgent necessity if the world is not to be reduced to barbarism.

Jawaharlal Nehru's reaction to the dropping of the atom bomb on Hiroshima and Nagasaki paralled Gandhiji's. He expressed his dismay at the "disastrous path modern civilisation is following," and added:

> Two great wars have brutalised humanity and made them think more and more in terms of violence. What progress, scientific, cultural and in human values we have made, is somehow twisted to the needs of violence.

In 1954, after the atom bomb had been overtaken by the hydrogen bomb, Jawaharlal Nehru, stressing that "the way of the atom bomb is not the way of peace or freedom," told the Indian Parliament:

> We have maintained that nuclear (including thermonuclear), chemical and biological (bacterial) knowledge and power should not be used to forge these weapons of mass destruction. We have advocated the prohibition of such weapons, by common consent, and immediately by agreement amongst those concerned, which latter is at present the only effective way to bring about their abandonment.

INDIRA GANDHI

On May 18, 1974, India carried out an underground nuclear explosion experiment at a depth of 100 metres in the Rajasthan desert. Prime Minister Indira Gandhi, clarifying to the Indian Parliament that "this experiment was part of the research and development work which the Atomic Energy Commission has been carrying on in pursuance of our national objective of harnessing atomic energy for peaceful purposes," said: "No technology is evil in itself; it is the use that nations make of technology which determines its character. India does not accept the principle of apartheid in any matter and technology is no exception."

> This view was in keeping with the point made by Mahatma Gandhi:

> That atomic energy, though harnessed by American scientists and army men for destructive purposes, may be utilised by other scientists for humanitarian purposes is undoubtedly within the realm of possibility.

And no one better realised this than Jawaharlal Nehru who initiated our programme of Atoms for Peace at the very dawn of independence:

> It is perfectly clear that atomic energy can be used for peaceful purposes, to the immense advantage of humanity. It may take some years before it can be used more or less economically (but) the use of atomic energy for peaceful purposes is far more important for a country like India whose power resources are limited than for an industrially advanced country.

In this light, having rejected the Non-Proliferation Treaty (NPT) of 1967 as an unequal treaty and undertaken the nuclear explosion at Pokhran in 1974, Prime Minister Indira Gandhi went on to lead the most significant disarmament initiative of the Eighties – the Five-Continent/Six-Nation Initiative – the tone for which was set by her immortal address to the Seventh Non-Aligned Summit in New Delhi at which she asked the key question:

Can there be peace alongside nuclear weapons?

Each day, each hour, the size and lethality of nuclear weapons increase. The hood of the cobra is spread. Humankind watches in frozen fear, hoping against hope that it will not strike. Never before has the earth faced so much death and danger. The destructive power contained in nuclear stockpiles can kill human life, indeed all life, many times over and might prevent its reappearance for ages to come. Terrifying is the vividness of such descriptions by scientists. Yet, some statesmen and strategists act as though, there is not much difference between these and earlier artillery pieces.

She then joined the Six-Nation appeal broadcast on May 22, 1984, which said:

The probability of nuclear holocaust increases and warning time decreases and the weapons become swifter, more accurate and more deadly. The rush towards global suicide must be stopped. We urge (a) halt (to) all testing, production and deployment of nuclear weapons and their delivery systems, to be immediately followed by substantial reductions in nuclear forces. We are convinced that it is possible to work out the details of an arrangement along these lines that takes into account the interests and concerns of all, and contains adequate measures for verification. This first step must be followed by a continuing programme of arms reductions leading to general and complete disarmament...

RAJIV GANDHI

On becoming prime minister, Rajiv Gandhi quickly established himself as an impassioned campaigner for universal nuclear disarmament, a campaign which reached its apothesis in the Action Plan he presented to the United Nations in 1988.

The Rajiv Gandhi Action Plan combines a practical roadmap towards universal, non-discriminatory nuclear disarmament (leading to general disarmament) and sustaining this by basing the world order on the principles of non-violence. The heart of the Action Plan lies in its emphasis of both a "nuclear weapon free" world and "non-violent world order" to sustain it.

Rajiv Gandhi's Action Plan was the culmination of forty years of intensive exploration of the road to nuclear disarmament. The heart of the Action Plan lay in the elimination of all nuclear weapons in three stages over a period of twenty-two years. Eighteen of these twenty-two years have passed with no progress even in the direction of the first stage. But abstracting from the specific time lines suggested in 1988, the three stages continue to remain valid. These are:

- First, a binding commitment by all nations to eliminate nuclear weapons in stages within a specific time-frame.
- Second, the participation of all nuclear weapon states in the process of nuclear disarmament, while ensuring that all other countries are also part of the process.
- Third, the demonstration of tangible progress at each stage towards the common goal.

The Action Plan further required that with a view to sustaining a world free of nuclear weapons, negotiations be undertaken to establish a comprehensive Global Security System under the aegis of the United Nations. To once again quote Rajiv Gandhi:

> When we eliminate nuclear weapons and reduce conventional forces to minimum defensive levels, the establishment of a non-violent world order is the only way of not relapsing into the irrationalities of the past. It is the only way of precluding the recommencement of an armaments spiral. Non-violence in international relations cannot be considered a Utopian goal. It is the only available basis for civilised survival, for the maintenance of peace through peaceful coexistence, for a new, just, equitable and democratic world order.

The arguments brought forward by Rajiv Gandhi in 1988 bear repetition even now, notwithstanding the enormous changes that have taken place in the international scenario over the last two decades and the fact that in the interim, India herself has moved from being a threshold nuclear power to a full-fledged NWS.

- First, now, as in 1988, nuclear war will mean the extinction of thousands of millions of human beings and the end of life as we know it on our planet Earth.
- Second, the relentless march of nuclear weapons technology renders ever more obsolete the pre-nuclear calculus of war and peace. As Robert S. McNamara has pointed out in a celebrated 2005 article in *Foreign Policy*, there are nearly 10,000 strategic offensive nuclear warheads in deployment worldwide, half of them by the United Stated with "the average US warhead (having) a destructive power 20 times that of the Hiroshima bomb."

In consequence, as Jawaharlal Nehru put it several decades earlier:

> These weapons, and the magnitude in which they will be employed, have

erased the difference between the capacity to inflict punishment and receiving the same; for the side that employs them is not immune from the effects of their own offence. It is a dangerous illusion to believe that nuclear weapons have brought us peace.

- Third, as Rajiv Gandhi told the UN:

There can be no iron-clad guarantee against the use of weapons of mass destruction. These have been used in the past. They could be used in the future. And, in this nuclear age, the insane logic of mutually assured destruction will ensure that nothing survives, that no one lives to tell the tale, that there is no one left to understand what went wrong and why.

- Fourth, as for the argument that since the consequences of nuclear war are widely known and well understood, therefore, nuclear war just cannot happen, it is again worthwhile to revisit Rajiv Gandhi's answer to the argument:

History is full of miscalculations. Perceptions are often totally at variance with reality. A madman's fantasy could unleash the end. An accident could trigger off a chain reaction which inexorably leads to doom.

The cautionary point made by McNamara in this regard is worth repeating:
The whole situation seems so bizarre as to be beyond belief. On any given day, as we go about our business, the President (of the United States) is prepared to make a decision within twenty minus that could launch one of the most devastating weapons in the world.

- Fifth, there is also little logic to the argument that as nuclear weapons have been invented, they, therefore, cannot be eliminated. There are several conventions already in operation relating to biological and chemical weapons of mass destruction. Only nuclear weapons remain outside the purview of a universal ban on weapons of mass destruction. The Action Plan signposts the stages by which nuclear disarmament too can be secured.

- Sixth, it remains as true today as it was in 1988, that, as Rajiv Gandhi put it:

There is nothing more dangerous than the illusion of limited nuclear war. It desensitises inhibitions about the use of nuclear weapons that could lead, in next to no time, to the outbreak of full-fledged nuclear war.

In 1988, the challenge of the Action Plan was essentially to doctrines of nuclear deterrence. That was at a time when two relatively well-matched "superpowers" were assuring their mutual survival by ensuring their mutual destruction. Now that hostility has give way to normalisation of relations between the two principal NWS, and all the self-certified NWS recognised by the NPT are promoting the best of relations among themselves, it is not so much the argument over the validity of deterrence doctrines as the need for the continued existence of weapons of mass destruction that takes centre-stage in the consideration of issues of nuclear disarmament.

Who are these weapons to be used against? Terrorists is one answer. But terrorist are non-state actors – and nuclear weapon are for use, if they are for use at all, only against hostile states or peoples. No one could suggest that the right response to a terrorist strike from a terrorist hideout could be a nuclear response. Indeed, the continued existence of large reserves of nuclear weaponry is the very treasure trove from which the terrorist hopes to filch his weapon of terror. Terrorism has, of course, to be fought but nuclear weapons can hardly be the weapon of choice.

The threat of nuclear proliferation will remain so long as an unequal world nuclear order legitimises the possession of such weapons in some hands, and those hands threaten the use of these weapons as a way of containing the threat of proliferation. The present juncture of a world without acute rivalries among the NWS is the right juncture at which to initiate an earnest dialogue under the aegis of the United Nations at the Conference on Disarmament based on the key concepts of the Rajiv Gandhi Action Plan that could lead to the realisation of the dashed hopes of the last two decades.

To this end, the Indian delegation to the 2006 session of the UN General Assembly circulated a Working Paper through which it reminded the international community that the Rajiv Gandhi Action Plan "provided a holistic framework seeking negotiations for a time-bound commitment for the complete elimination of nuclear weapons to usher in a world free of nuclear weapons and rooted in non-violence." With this in view, the Working Paper calls on the international community to build a consensus that strengthens the ability of the international community to initiate concrete steps towards achieving the goal of nuclear disarmament based on the following elements:

- Reaffirmation of the unequivocal commitment of all nuclear weapon states to the goal of complete elimination of nuclear weapons.
- Reduction of the salience of nuclear weapons in security doctrines.
- Taking into account the global reach and menace of nuclear weapons, adoption of measures by nuclear weapon states to reduce nuclear danger, including the risks of accidental war, de-alerting of nuclear weapons to prevent unintentional and accidental use of nuclear weapons.
- Negotiations of a global agreement among nuclear weapon states on "no first use" of nuclear weapons.
- Negotiations of a universal and legally-binding agreement on non-use of nuclear weapons against non-nuclear weapon states.

- Negotiation of a convention on the complete prohibition on the use or threat of use of nuclear weapons.
- Negotiation of a Nuclear Weapons Conventions prohibiting the development, production, stockpiling and use of nuclear weapons, and on their destruction, leading to the global non-discriminatory and verifiable elimination of nuclear weapons with a specified time-frame.

The participants in this international conference convened to celebrate the centenary of Mahatma Gandhi's *Satyagraha* are invited to lend their voice to a renewal of the momentum towards universal non-discriminatory nuclear disarmament by endorsing the Working Paper to open negotiations aimed at securing an intelligent consensus which might yet save humanity. ■

Valedictory Address

☐ **M. HAMID ANSARI**

TO DELIVER a valedictory address is to begin with twin apprehensions: that whatever is to be said has been said, or that what is going to be said may be outrageously at variance with what has been said! In either case, the speaker would be deprived of what Dr. Samuel Johnson termed "frigid tranquillity".

For being in this interesting position, I have to thank an old and dear friend!!

Any discussion necessitates clarity about concepts. In terms of the theory of statecraft, war has always been considered an instrument of policy. Pursuit of war necessitates weapons, defined as tools to gain advantage over an adversary. Improvement in the quality of weapons, and invention of new ones, is a logical outcome of the human trait to seek excellence, for success in subduing a political and military adversary by inflicting unacceptable damage.

The impulse to invent weapons of mass destruction, including nuclear weapons, was part of this process. Each new weapon system also propelled assessment of its implications in tactical and strategic terms. Both processes were accelerated in the second half of the 20th century.

Every invention, apart from its novelty, has to prove its utility. Mass destruction in the Spanish civil war was vividly depicted by Picasso's Guernica; less than a decade later, it was typified by London and Dresden. In the case of the nuclear weapon, the utility was brutally demonstrated at Hiroshima and Nagasaki. The scale of destruction there propelled consideration of the implications of the new weapon.

An early recognition came in the shape of the Baruch Plan of June 1946. It was rejected by the Soviet Union for reasons that were evident. In 1948, General Omar Bradley told an American audience that "the only way to win an atomic war is to make certain it never starts."

None appreciated the implications better than the scientists. In July 1955 the signatories of the Russell-Einstein Manifesto spoke "not as members of this or that nation, continent, or creed, but as human beings, members of the species man, whose continued existence is in doubt."

The venue of this conference is important; so is its timing. India has been an ardent advocate of prohibition on the production and use of nuclear weapons. Jawaharlal Nehru in 1954 spoke of the fear that "would grow and grip nations and peoples and each would try frantically to get this new weapon or some adequate protection from it." Prime ministers of India proposed prohibition in 1978 and again in 1982. In 1988, Rajiv Gandhi sought "not a marginal adjustment in the machinery of nuclear confrontation, nor a partial or temporary scaling down of the arms race," but "a world which is rid of nuclear weapons." His Action Plan for a World Free of Nuclear Weapons was comprehensive in its scope, passionate in its appeal, and clinical in its reasoning and analysis.

The idea was considered utopian. Despite this, and in the initial euphoria at the end of the Cold War, the global momentum for a less unpleasant world led to the conclusion of the universal and non-discriminatory Chemical Weapons Convention, with its intricate and intrusive verification mechanism. Significantly, however, the argument for outlawing it was not extended to nuclear weapons.

In regard to matters nuclear, the world has witnessed changes over the past decade and a half. This audience is knowledgeable about it. India herself has emerged as a nuclear weapon state. On one side it is argued that the imperative of realism leaves no option but to accept the reality. On the other, those distressed over the fraying of world order and apprehensive of the "normative cost of silence" advocate a more assertive approach. "This is a time," writes Professor Richard Falk in his recent book *The Costs of War*, "when realism and idealism are increasingly fused in their call for a future world order based on law and justice, but this cannot be made to happen without the engagement of the peoples of the earth acting as detribalized citizens without borders."

Three questions arise:

- Is the logic of 'realism' unassailable?
- Does it hold good for the world of tomorrow?
- Has the argument for disarmament, and particularly for nuclear disarmament, ceased to be relevant for the survival of the human species?

The case for the possession of nuclear weapons needs to be assessed in strategic, legal, political, financial, developmental and environmental terms. This would unavoidably widen the ambit of discourse.

In the first place and according to the Federation of American Scientists, the global stockpile of nuclear warheads today remains at more than 20,000. Of these, more than 10,000 warheads are considered operational, of which a couple of thousand are on high alert, ready for use on short notice. The approach is premised on the doctrine of deterrence; the latter, however, remains inherently unstable, prone to human error or folly; the probability of the annihilation of the human race through the use of these weapons, thus, remains high, and must be considered unacceptable.

Secondly, nuclear armament ends up being, in its implications, anti-poor and anti-development. Stephen Schwartz, in his 1998 book *Atomic Audit* on the comprehensive cost of the US nuclear weapons programme, has estimated that the US spent around $6 trillion in total. The arms race led the former Soviet Union to the point of exhaustion and disintegration. The resource drain of other nuclear weapon states would be equally high in proportionate terms. This level of spending by nuclear weapon states cannot but deny national resources for developmental or other purposes for public welfare.

Thirdly, the development, production, stockpiling and use of nuclear weapons result in immense, irreversible and unforeseen damage to the environment. As far back as 1987, the Report of the World Commission on Environment and Development (known as the Brundtland Report), affirmed that "among the dangers facing the environment, the possibility of nuclear war is undoubtedly the gravest". It noted that the "whole notion of security as traditionally understood in terms of political and military threats to national sovereignty must be expanded to include the growing impacts of environmental stress"; it concluded that "there are no military solutions to 'environmental insecurity'."

Fourthly, there have been fundamental changes in the nature of conflict and of the structure of international relations. Conflict in the post-Cold War era has acquired new characteristics: it is not classical inter-state conflict; it is fuelled by identity based factors and issues of economic and social justice; and there is a drastic increase in the role of non-state actors. Weapons of mass destruction that were fashioned for inter-state conflict and their associated strategic deterrence doctrines premised on state behaviour have little relevance for the new reality.

The case for possession and use of nuclear weapons stands dented and lends credence to the need to rethink its fundamentals. Any endeavour on this basis must necessarily be rooted in legality and morality and be capable of demonstrating the advantages emanating from it.

How is this elusive goal to be attained? In exploring options, we need to remember that the community of nations has put in place agreements to outlaw chemical and biological weapons.

The question of legality poses problems. On a reference from the UN General Assembly on "threat or use of nuclear weapons," the International Court of Justice (ICJ) gave an Advisory Opinion in July 1996. It decided that in customary or conventional international law there is neither an authorisation nor a prohibition of the threat or use of nuclear weapons. While it opined that such a threat or use would be contrary to the rules of international law applicable to armed conflicts, it noted that the current state of international law does not permit the Court to conclude definitively whether such threat or use would be lawful or unlawful in an extreme circumstance of self-defence, in which the very survival of a state would be at stake.

The distinguished audience here is cognisant of the UN General Assembly resolutions passed each year by a large majority reaffirming that "any use of nuclear weapons would be a violation of the Charter of the United Nations and a crime against humanity" as declared in its Resolution 1653 (XVI) of November 24, 1961.

This reveals the desire of a very large section of the international community to move forward along the road to complete nuclear disarmament. On the other hand, we have the annual reaffirmation of the Chapter VII Security Council Resolution 1540 of 2004 stating that proliferation of nuclear weapons "constitutes a threat to international peace and security".

Put together, we get two sets of assertions:

1. use of nuclear weapons is a crime against humanity;

2. proliferation of nuclear weapons is a threat against international peace and security.

Between the two ends of this spectrum, falls the question of production, possession and threat to use of nuclear weapons. It is an irony of *realpolitik* that these have so far not been perceived to constitute a threat to international peace and security.

The ICJ addressed but did not resolve a critical question: Would a higher priority be accorded to the survival of the state if the survival of humanity itself were at stake?

A Dissenting Opinion summed up the legal dilemma: "The case as a whole presents an unparalled tension between State practice and legal principle." "When it comes to the supreme interests of State", it noted, "the Court discards the legal progress of the Twentieth Century, puts aside the provisions of the United Nations Charter of which it is the principal judicial organ, and proclaims, in terms redolent of *Realpolitik*, its ambivalence about the most important provisions of modern international law." This impasse was reflected most recently in the Report of the UN Disarmament Commission on April 25, 2008 at the end of its Three-Year Cycle of Deliberations. Releasing the Report, the chairman of the commission said that even set against the relatively low expectations, the results were meagre. "There was a stark contrast between the state of the world and the cooperation of the United Nations Member States in the Commission. Therefore, the credibility question is inescapable, and in time, each and every one of us should be able to answer it."

The only way to resolve the impasse is to do it on a different plane. The modern state system is premised on the model emanating from the Peace of Westphalia. The reality of the sovereign state today, however, is very different from its theory. In 1991, Javier Perez de Cuellar had called upon the international community to help develop a "new concept, one which marries law and morality."

Such an effort of bringing together law and morality would help initiate the process of resolving the dilemma highlighted by the ICJ in its Advisory Opinion. The process would then take us back to the Russell-Einstein Manifesto's focus on the human being:

> Remember your humanity, and forget the rest. If you can do so, the way lies open to a new paradise; if you cannot, there lies before you the risk of universal death.

To transform vision into reality, a plan and a timetable on the pattern of the Rajiv Gandhi Action Plan would be essential. We have seen that, hitherto, nuclear disarmament has become almost synonymous with nuclear non-proliferation. A change would be possible only through such an Action Plan.

For much too long, ladies and gentlemen, the question of disarmament has remained in the exclusive domain of states and their experts. Is it not time now to open a window or two to let in the fresh breeze of global public opinion? We are aware of the beneficial results produced by such an approach in other areas that transcend state sovereignty.

Given the immobility of the current disarmament process, a new methodology may be worth a try. ∎

Beyond Deterrence

☐ **JONATHAN GRANOFF**

MR. PRIME MINISTER, Ambassador Duarte, Air Commodore Singh, Mr. Chairman, Excellencies, the Centre for Strategic and International Studies and the Indian Council for World Affairs, please accept my gratitude for the privilege of addressing this important gathering. I would like to honour the powerful words of the sage Swami Vivekananda at the first Council for a Parliament of World Religions in Chicago in 1893 where he began his speech, "Sisters and Brothers of America." Please permit me to say with heartfelt enthusiasm, "Sisters and Brothers of India."

It might appear that we have come from great distances to be here, but as never before the world is one neighbourhood and we are all living very close to one another. It is interesting how a suburban American like myself has come to appreciate this truth. Near my home, I buy my gas from a Russian concession. The bank for my country, acknowledging the reality of current debt and currency resources, is your neighbour, China. When my phone or computer needs service, I consult a technical expert in Bangalore or Mumbai. Our lives are so deeply interconnected, and our destinies, one.

What an honour to be in a nation which can serve the world as an example of how great diversity can lead to cultural enrichment, democratic participation, and grow united in common purposes.

For those who appreciate the mystery of the life of civilisation, India represents more than a great nation. Its name resonates in the heart as a cultural spiritual treasury for all humanity. The Indus Valley civilisation reminds us of Valmiki, Tulsidas, Mahavir, Buddha, Khwaja Muinuddeen Chishti, Nizamuddin, Guru Nanak, Asoka, Bawa Muhaiyaddeen, Ramana Maharshi, Rama Krishna, Mirabai, Mahatma Gandhi, and so many other luminaries, and, in this gathering today, the political leadership of Rajiv Gandhi. Every name reminds us of leadership in the practice of our core humanity. And this remembrance raises our vision to a place where nuclear weapons should have no acceptable position.

To use nuclear weapons against a nuclear weapon state is suicidal and against a non-nuclear weapon state, genocidal and patently unacceptably immoral. They have no possible use against terrorists.

During the Cold War, nuclear deterrence theory reached its apex as a framework paradoxically designed to ensure non-use. However, the paradox of nuclear deterrence is that for it to work, the threat of use must be perceived as real. Dancing on the edge of catastrophe must become accepted as a daily reality. For this dangerous dance to be sustained, one must find a partner who knows and can be relied upon to follow its steps. To fail is fatal.

The USSR and the US did not dance cheek to cheek and yet we should remember that only good luck or God's grace has allowed us to be alive here today, all too closely having avoided the *danse macabre*.

Deterrence is too dangerous. Even under the best of circumstances, mistakes can be made. General Lee Butler was US commander of Strategic Nuclear Forces, with the day-to-day responsibility for operations, discipline, training of tens of thousands of crew members, the systems that they operated and the warheads those systems were designed to deliver. He said that after he studied deeply into the history of the incidents and the accidents of the nuclear age as they had been recorded by the US and USSR "…it is more chilling than anything you can imagine." He recounted, "Missiles that blew up in their silos and ejected their nuclear warheads outside of the confines of the silo. B-52 aircraft that collided with tankers and scattered nuclear weapons across the coast and into the offshore seas of Spain. A B-52 bomber with nuclear weapons aboard that crashed in North Carolina, and on investigation it was discovered that on one of those weapons, 6 of the 7 safety devices that prevent a nuclear explosion had failed as a result of the crash. There are dozens of such incidents. Nuclear missile-laden submarines that experienced catastrophic accidents and now lie at the bottom of the ocean." The Cuban Missile Crisis gave the world 13 days to reach safety. How much time is enough to rectify human or mechanical error?

The margin of error with close neighbours is negligible and any error unforgiving. What would happen if a playful computer hacker crossed a line where good humour became Armageddon? Would the motive matter?

India has a deep existential interest in eliminating nuclear weapons rapidly. Moreover, amongst the nuclear weapon states, it alone has consistently and correctly set forth coherent arguments for abolition. India has highlighted that nuclear apartheid is practically and morally unacceptable and unsustainable. Gaining the status of the privileged does not change the correctness of the analysis, but it helps provide India with the credibility to lead in a movement for a nuclear weapon free world. Sometimes it takes another piece of wood to take out a splinter.

Will India be alone in this effort? When the likes of George Schultz, Henry Kissinger, William Perry and Sam Nunn pen op-eds in the *Wall Street Journal* emphasising that nuclear deterrence is inadequate to obtain security, and efforts at non-proliferation cannot succeed without a clear commitment and path to obtain nuclear weapons abolition, none can any longer assert that abolition is either impractical or anti-American. It took sixty years for such "realists" to reach this conclusion. We thought that US nuclear superiority would bring security but the arms race with the Soviet Union proved us wrong. The nuclear weapon states thought they could preach abstinence to South Asia while sitting on bar stools—India and Pakistan proved them wrong. If we all persist in the illusion that incoherence will bring stability, we do not know who will prove us wrong next. And certainly Iraq has demonstrated that the approach to contain nuclear threats through counter-proliferation is hardly an answer.

I am pleased to share that I am part of a working group that arose under the leadership of President Gorbachev and George Schultz at a conference at Harvard University's Belfer Centre to craft recommendations for the next presidents of Russia and the United States to move toward the elimination of nuclear weapons. It should be clear that many more oarsmen will be coming on board to join such good efforts in the near future.

The choice is between attempting to maintain the status quo of a two-tiered security system which will result in ongoing creeping proliferation, clearly a route to hell fire, or global legally verifiable and enforceable elimination. This would be the fulfillment of India's moral vision for a better world and a gift to humanity in the name of all that is good as well as that of a good and great man, Rajiv Gandhi.

A nuclear free world is a world where the principle of equity is honoured. Equity breeds stability, and inequity, instability.

But there is an even deeper reason, a universal reason, aptly described in the ancient epic, the *Ramayana*. In his dissenting opinion in the International Court of Justice's famous Advisory Opinion on the legality of the threat or use of nuclear weapons, Judge Weeramantry summarised this wondrous story in relevant part. During the war between Rama and Ravana – rulers respectively in India and Sri Lanka – "a weapon of war become available to Rama's half-brother, Lakshmana, which could 'destroy the entire race of the enemy, including those who could not bear arms.' Rama advised Lakshmana that the weapon could not be used in war 'because such devastation *en masse* was forbidden by the ancient laws of war, even though Ravana was fighting an unjust war with an unrighteous motive'." Such prohibitions were ancient several thousand years ago. This was truth then and it remains true now. Why should we find the use of the plague so abhorrent as a weapon and accept the threatening devastation of nuclear devices? It is the indiscriminant and devastating effects of weapons of mass destruction which present an unacceptable threat.

India cannot eliminate unacceptable threats to its security by force. Pakistan will never disarm without India's disarming and India will certainly not disarm without a similar gesture from China. Will China do so without Russia, and Russia without the US? Thus, we come back to the original logic of India — let us follow a clear, direct route to nuclear disarmament that is non-discriminatory, comprehensive and universal.

How can this be best done now? It is time for stepping up the process. I, thus, suggest that India convene at the earliest possible time, a Nuclear Weapons Convention Preparatory Conference to examine the conditions to obtain a Nuclear Weapons Convention and forge a path to its achievement.

India is uniquely positioned to advance this route. It has the political arguments, the national interest, the cultural and moral calling, and historical moment. The world needs the compass point of leadership. Where can this be found? Is it presumptuous for an American to call upon India to seize such a flag for us all? If nuclear weapons teach us anything good, it is a truth known for millennium by the wise everywhere: humanity is one family. A catastrophe for any nation with these devices is a catastrophe for all. And, of course, the leadership of one will be of service and benefit to us all. The ancient culture and wisdom of India identifies it as that leader today. ■

Nuclear Weapons and International Terrorism

☐ **RAJESH RAJAGOPALAN**

NUCLEAR terrorism has become a significant concern over the last decade. Though there have been concerns about nuclear terrorism even prior to the 9/11 attacks, that attack has increased attention to the possibility of terrorists acquiring nuclear weapons.[1] Though some of the more alarmist analyses has now given way to more tempered considerations of the threat, there is still considerable worry about the possibility of terrorists arming themselves with nuclear weapons.[2] But, as we consider the threat posed by terrorists armed with nuclear weapons, we should also consider why terrorists have not resorted to nuclear weapons until now. Understanding the utility of nuclear weapons, as well as their limitations, will allow us to get a better handle on the problem of nuclear terrorism.

In this paper, I argue that the threat of nuclear terrorism has been limited until now by the general lack of utility of nuclear weapons in promoting any conceivable political goals that terrorist groups might have. This limitation is primarily the consequence of the nuclear 'taboo' – the norm against the use of nuclear weapons that has held strong since the atomic bombing of Hiroshima and Nagasaki. Though this does not ensure that no terrorist group will resort to nuclear weapons in the future, the likelihood of such use will remain limited as long as that taboo holds.

This paper is divided into five sections. In the first section, I look at the prospects of terrorists acquiring nuclear weapons. The second section examines why nuclear weapons have not so far been used by some of the more capable terrorist groups. The next section examines some of the counter-measures that can be adopted against the threat of nuclear terrorism while the fourth section looks at a related policy option more closely: whether nuclear weapons can be used to deter terrorists, specifically those armed with nuclear weapons. The final substantive section looks at the unthinkable – policy options in the aftermath of a nuclear terrorist incident.

CAN TERRORISTS ACQUIRE NUCLEAR WEAPONS?

There are a number of different ways in which terrorists can acquire nuclear weapons.[3] They could build, buy or steal one, or they could take control of a nuclear facility. All of these measures are possible, but, clearly, they represent different levels of difficulty for any terrorist group.

Building a nuclear weapon is possibly the most difficult of the various measures for acquiring nuclear weapons.[4] The difficulties that even national governments have had in building nuclear weapons suggest that this will be a very difficult, even impossibly difficult, for terrorist groups. Many states have spent huge quantities of their national wealth in the hope of building nuclear weapons but have not succeeded. Many of these were states that had the determination and the wealth to carry through such projects, but they found it hard to bring the material, the personnel, and the technology together to successfully complete such projects. Even those states (other than the nuclear five) that have built such weapons have done so with great difficulty and after considerable effort. What this suggests is that terrorists groups, with far lesser capability in every measure, will not find this task easy.

Two caveats must be noted, however. States that seek to build nuclear weapons are usually seeking to build an arsenal, rather than a single weapon, and typically seek to create an infrastructure for a continuing weapons programme. Terrorist groups that want to fashion a single device will not require such infrastructure and this could reduce their difficulties. Second, though states are more capable than terrorist groups, they are also under greater scrutiny by the international community. This is particularly true of states that either have some demonstrated technical capability or those that have expressed a desire to have such weapons. Such scrutiny makes it difficult for states to pursue nuclear weapons without being detected. Despite these caveats, however, the difficulties that terrorist groups face in building their nuclear weapon are so great that we can rule out the prospect with some confidence.

Buying or stealing a nuclear weapon is a less serious challenge for a financially well-endowed terrorist group.[5] Nevertheless, the difficulties here are also considerable. Precisely because of their lethality, nuclear weapons are well-guarded and protected with electronic locks that require considerable effort to circumvent. In addition, nuclear weapons in many countries are kept de-mated, which makes it very difficult to acquire an entire weapon even if there is a successful heist.

NON-WEAPON THREATS

Terrorists could also decide to use variants rather than nuclear weapons to create panic. These include the use of radiological weapons (or, more accurately, a radiological dispersion device – RDD) or attacking a nuclear facility in the hope of spreading radiological contamination from the facility. Reportedly, the Al Qaeda had planned to crash the passenger jets it hijacked as part of the 9/11 plot on a nuclear facility.

Both of these are far easier to achieve for any terrorist group, but they are also far less effective. An RDD attack would be most effective in creating panic if those attacked had no prior information about RDD weapons and mistook an RDD attack for a nuclear attack. But over the last several years, there has been an intense debate about such threats that has had at least one positive effect. Because information about such weapons is now widely available, it reduces one of the potent effects of such weapons: panic. Such weapons are not capable of killing many people directly, but they might have a psychological effect that is still potent.

Attacking a nuclear facility in the hope of spreading contamination remains one of the most serious nuclear threats posed by terrorists. But here too, there has been widespread debate, tightening of physical security around such facilities but also some disagreement about how vulnerable facilities such as nuclear power plants as well as other facilities such as those where spent fuel is stored are to attack. Overall, however, I would rate the risks of such action as higher than an attack with nuclear weapons, but the effectivesness of such attacks would be somewhat low.

THE UNATTRACTIVENESS OF NUCLEAR TERRORISM

Despite the difficulties of acquiring nuclear weapons, some of the more effective established terrorist groups, including the Liberation Tigers of Tamil Eelam (LTTE), Hamas, and Hezbollah are sufficiently well-organised and supported that it is within their technical, financial and organisational capabilities to have acquired nuclear weapons or RDDs. Nevertheless, there is little indication that such groups have ever sought any variety of nuclear weapons. If the reason for their failure to pursue nuclear weapons is not lack of capability, then we have to assume that there are good political reasons for such reticence.

One reason for the unattractiveness of nuclear weapons as a terrorist tool is that any use of nuclear weapons would immediately make both the group that conducts such acts, as well as the cause that that group espouses, illegitimate, as noted by Kenneth Waltz.[6] A group that uses such weapons will quickly find itself isolated and probably the target of joint military action by several states and not just the state that they target. If such groups had been sponsored or supported by another state, their sponsor would also abandon them, not wishing to be tarred by association with such groups.

But this raises a more fundamental issue. Why is it that such uses of nuclear weapons are illegitimate? The answer lies in the larger illegitimacy of the use of nuclear weapons itself. The nuclear taboo, as some scholars have somewhat inaccurately termed it, is not a recent phenomenon.[7] In the 1960s, Thomas Schelling noted this as "the tradition of non-use of nuclear weapons."[8] The strength of this taboo can be witnessed in the fact that many nuclear armed countries have preferred to lose wars rather than use their nuclear arms to stave off defeat. This nuclear taboo, though primarily a normative constraint, is a fairly strong restraint even for non-state groups when they contemplate their choices. It should also be noted that such taboo extends also to other weapons, including chemical weapons, which makes their use difficult, even under trying circumstances.[9]

CATASTROPHIC TERROR

Would such normative restraint work on groups such as the Al Qaeda or other groups in the age of "catastrophic terrorism"? Possibly not. Groups such as these do not have political goals that can be the subject of political negotiations; these groups, in this respect at least, are not 'political'. And precisely because of the apolitical nature of these groups, it is likely that if they could, they would use nuclear weapons. With such groups, prevention is the only cure.[10] International cooperation to prevent such groups from acquiring such tools is essential.

Nevertheless, we need to keep this threat in perspective. For one, there are not very many of these groups. If we focus not on all such groups but on those that have the organisational and material capacities to actually undertake such missions, the list dwindles even further. Outside of the Al Qaeda, it is difficult to think of any group that has the motivation, support, skill and resources to conduct such acts. Even the Al Qaeda may no longer have such capability, after years of suffering attrition from the US and other states and after having lost its safe-haven in Afghanistan. Other factors also need to be kept in mind: acquisition of nuclear weapons would force such groups to expand their tightly-knit cellular organisational structure, making them more vulnerable. In addition, under international pressure, many states that might have aided the Al Qaeda or at least looked the other way, can no longer be expected to do so, increasing the risks of any such operation.

None of this should be taken to mean that the Al Qaeda – or another such group – will not succeed in such missions. But such groups face significant and growing challenges that limit their capacity to engage in nuclear terrorism.

COUNTERING NUCLEAR TERRORISM

The conclusion to be drawn from the previous two sections is that though nuclear terrorism is not impossible, especially for groups engaging in catastrophic terrorism, most 'political' groups will abjure nuclear weapons, even though some of these groups are probably more capable of acquiring them. Nevertheless, additional measures can be taken to further reduce the threat from nuclear terrorism.

The most obvious part of such additional measures is greater national and international intelligence and police cooperation to tackle such threats before they arise. A number of measures, including the International Convention for the Suppression of Acts of Nuclear Terrorism (2005), have already been undertaken. Additional measures have been undertaken through the International Atomic Energy Agency (IAEA) in the areas of securing fissile materials. These measures definitely go a long way in increasing the difficulty that terrorists would face in acquiring nuclear material. These measures are particularly useful in preventing groups espousing catastrophic terrorism from achieving their ends.

But we also need to pay greater attention to some of the more normative elements associated with the nuclear weapons use taboo. One of the reasons why the threat of nuclear terror has remained at a manageable level is because, as I argued earlier, many of the more capable terrorist groups have not sought nuclear weapons because of the taboo associated with their use. The problem of nuclear terrorism would have been infinitely worse if the use of nuclear weapons had acquired any measure of legitimacy. The key question then is how to sustain and strengthen the nuclear weapons use taboo. Clearly, nuclear-armed states have the primary responsibility here and their actions are the key.

The complete elimination of nuclear weapons would, of course, be nice not only for the usually listed reasons, but also because it would reduce the threat of nuclear terrorism considerably though it would not eliminate it. The groups that we worry about the most – those that practise, or hope to conduct catastrophic terrorism – are also the ones that are least likely to be influenced by such measures. So even if a Nuclear Weapons Convention were to be successful, it is unlikely that it would eliminate the risk of nuclear terrorism. But such a global ban could increase the difficulties faced by terrorist groups who will then be forced to actually build a nuclear weapon, a far more challenging prospect than buying or stealing a built weapon.

Nuclear disarmament is also a somewhat distant and uncertain prospect. Though not necessarily contradictory, more immediate measures could be adopted that could have greater potential for strengthening the nuclear weapons use taboo. Two immediate measures that can be suggested are de-alerting and de-mating of nuclear weapons and a no first use (NFU) agreement. Minimally, states that possess nuclear weapons can take measures to reduce their reliance on such weapons.

De-alerting, and more importantly, de-mating, can help significantly in reducing the probability of terrorists acquiring nuclear weapons either through subverting personnel who control nuclear weapons or by stealing such weapons. De-mating includes separation of nuclear warheads from their delivery systems as well as separation of the nuclear fissile core from the rest of the warhead. The weapon components are typically held by different agencies, possibly in different locations. Such separation would present great difficulties for anyone wanting to acquire illegal control over the weapon because it would require multiple operations to steal or acquire control over different components. This would be a much harder task than stealing a single weapon. Though these weapon components would be brought together periodically during military exercises, it would, nevertheless, present much greater difficulties for terrorist groups to acquire control over these weapons.

In addition to helping in securing physical control over the weapons, de-alerting and de-mating of weapons would also send out a powerful message about the illegitimacy of these weapons for normal military use, thus, strengthening the existing nuclear weapons use taboo. In essence, such measures would indicate that all countries that possess nuclear weapons accept that nuclear weapons are in a special league and that their use can be contemplated only in the most dire of circumstances. As of today, several nuclear-armed states have opposed such initiatives at the UN, while insisting that their weapons are not on hair-trigger alert. Analysts such as Bruce Blair have pointed out, in response, that even if such weapons are not on hair-trigger alert, they are on a very high state of alert.[11]

An NFU pledge would also have similar effects. In essence, by accepting NFU, states that possess nuclear weapons can further emphasise that they would not consider nuclear weapons to be legitimate tools of warfare. An NFU pledge is also a recognition of the reality that whatever peace-time nuclear doctrines and strategies are adopted, in a crisis, most political decision-makers tend to be extremely cautious about contemplating policy options that raise the possibility of nuclear escalation. Throughout the Cold War, despite declared muscular nuclear strategies, backed up by similar force postures and command and control arrangements, both the US and the Soviet Union exercised great caution to ensure that they never stepped on the escalation ladder. An NFU pledge only recognises this reality, and in the bargain, it not only reduces the risk that accompanies any nuclear arsenal but also reduces the threat of nuclear terrorism by further delegitimising these weapons.

If states possessing nuclear arms can reduce the risk of nuclear terrorism by these measures, it is also in their power to actually increase that risk. Unfortunately, nuclear strategies adopted by various states, including India, over the last decade do just this – increase the risks. They do so by adopting nuclear doctrines that propose to enlarge the circumstances under which they would consider the use of nuclear weapons. Contemplating nuclear weapons use against terrorists, or in response to chemical and biological weapons attacks, or as a substitute for weak conventional forces, all expand the potential conditions under which nuclear weapons use could be considered. Even if such policies do not require states to actually use nuclear weapons in such contingencies, they raise the risk of what is called the "commitment trap" – the possibility that political decision-makers would be trapped by peace-time policy commitments into adopting options that they otherwise would not have chosen.[12] But more important than the consequences of such steps for actual war-time policy option is the impact that such policy declarations have on the perceived legitimacy of nuclear weapons. Such declarations, by suggesting that nuclear weapons use could be considered under many circumstances rather than purely to deter and retaliate against the use of other nuclear weapons, increase the legitimacy accorded to such weapons. While such policies will not have an immediate impact in making nuclear weapons use legitimate or undermining the nuclear taboo, they can have such effects over the long term.

CONCLUSIONS

In essence, then, though the risks of nuclear terrorism remain, the threats are somewhat limited. Only a few groups have sought such weapons, and most are not capable of acquiring them. The most effective and capable terrorist groups have not sought these weapons because they would not be able to use them effectively to advance their political agenda. This is the result of the illegitimacy of the use of nuclear weapons. In seeking to prevent nuclear terrorism, states can help in strengthening the nuclear taboo by eschewing aggressive doctrines and high-alert states for their nuclear arsenals, and by moving towards a de-alerted, de-mated posture, with an NFU policy. Any dilution of the nuclear taboo could make those groups that now do not seek nuclear weapons to also consider them, making the challenge of nuclear terrorism far greater. ■

NOTES

1. For an earlier view of nuclear terrorism, see *Report of the International Task Force on the Prevention of Nuclear Terrorism* (Washington D.C.: The Nuclear Control Institute, 1986)

2. Graham Allison, *Nuclear Terrorism: The Ultimate Preventable Catastrophe* (New York:

Times Books, 2004).

3. Bruce G. Blair, "What If the Terrorists Go Nuclear?" available at <http://www.cdi.org/terrorism/nuclear.cfm>

4. For a dated but still informative analysis of the technical issues involved, see Carson Mark, Theodore Taylor, Eugene Eyster, William Maraman, and Jacob Weschler, "Can Terrorists Build Nuclear Weapons?" Paper prepared for the International Task Force on the Prevention of Nuclear Terrorism, 1986, available at <http://www.nci.org/k-m/makeab.htm>

5. On measures to increase security against such threats, see M. Bunn and G. Bunn, "Reducing the Threat of Nuclear Theft and Sabotage," at <http://www.iaea.org/NewsCenter/Features/Nuclear_Terrorism/bunn02.pdf >

6. Kenneth N. Waltz, "Waltz Responds to Sagan," in Scott D. Sagan and Kenneth N. Waltz, *The Spread of Nuclear Weapons: A Debate Renewed* (New York: W.W.Norton & Company, 2003), pp. 126-30.

7. On how the nuclear taboo has prevented the US from using nuclear weapons, see Nina Tannenwald, "The Nuclear Taboo: The United States and Normative Basis of Nuclear Nonuse," *International Organization*, 53:3, Summer 1999, pp. 433-468.

8. Thomas Schelling, *The Strategy of Conflict* (Cambridge, Mass.: Harvard University Press, 1960)

9. Richard Price, *The Chemical Weapons Taboo* (Ithaca, New York: Cornell University Press, 1997)

10. Scott D. Sagan, "Sagan Responds to Waltz," in Sagan and Waltz, n.6, pp. 159-164.

11. Wade Boese, "Nuclear Weapons Alert Status Debated," *Arms Control Today*, December 2007.

12. Scott D. Sagan, "The Commitment Trap: Why the United States Should Not Use Nuclear Threats to Deter Biological and Chemical Weapons Attacks," *International Security*, 24:4, Spring 2000, pp. 85-115.

The Leaders Must Be Led

Responsibilities of the US and Russia in Leading the Way to Nuclear Disarmament

☐ **DOUGLAS ROCHE**

THERE is no doubt that the United States and Russia, possessing 95 per cent of the world's 25,000 nuclear weapons, have the chief responsibility in leading the way to nuclear disarmament. But far from doing this, they are standing in the way of genuine progress. In fact, they themselves must be led by other states and an energised civil society to recognise that, in addition to their moral and legal responsibilities, it is in their interests to move toward a world free of nuclear weapons. In short, the nuclear leaders must be led. This paradox will require creativity, ingenuity and, not least, determination in order to resolve the nuclear weapons dilemma. I believe the task can be achieved, particularly if India plays a leading role.

Both the US and Russia would have us believe that they are in full compliance with their obligations under Article VI of the Non-Proliferation Treaty (NPT). They claim to have made "unprecedented progress" since the end of the Cold War in the field of nuclear disarmament. Certainly, eliminating some 40,000 nuclear weapons from the Cold War peak of 65,000 is no small feat. Were they to continue to eliminate nuclear weapons at this rate, we could cheer. But the reductions so far have been a smokescreen for their stated intentions to maintain nuclear arsenals far into the 21st century. While taking credit for chucking superfluous nuclear weapons, both countries continue modernising programmes that will enable them to inflict immense and catastrophic destruction with fewer nuclear weapons.

The Nuclear Posture Review, conducted by the Bush Administration in 2001, established expansive plans to "revitalise" US nuclear forces and all the elements that support them within a new triad of capacities combining nuclear and conventional offensive strikes with missile defences and nuclear weapons infrastructure. Under the subsequent post-9/11 National Security Strategy, the Administration said it would take "anticipatory action" (a euphemism for preemptive strikes) against enemies of the United States, and has not ruled out using nuclear weapons, which remain a cornerstone of US national security policy.

Although Congress has blocked funding for the Reliable Replacement Warhead programme, it left intact funding for other programmes that functionally include ongoing nuclear weapons research and development and provide for the training of new nuclear weapons scientists and engineers. In fact, Congress has authorised the Administration "to develop and submit to the Congress a comprehensive nuclear weapons strategy for the 21st century." This action is in keeping with the call by the Commander of US Strategic Command, General Kevin Chilton, for the US to maintain nuclear weapons for the remainder of the 21st century. Both the Administration and Congress continue to anticipate the need to revitalise the present nuclear weapons complex. This approach renders meaningless the disarmament objective implicit in reductions of current numbers of weapons.

US determination to maintain nuclear weapons in the 21st century led former Russian President Vladimir Putin to announce in 2004 that Russia is carrying out research and missile tests of state-of-the art nuclear missile systems. Russia, he said, would "continue to build up firmly and insistently our armed forces including the nuclear component." The Russians, far from lowering the political value of nuclear weapons, have elevated them in their posture because of their conventional arms weakness. Nuclear dismantlements are offset by the continued deployment of newer, more capable road-mobile Topol-M missiles. A new Borey-class of nuclear ballistic missile submarines is underway.

The dangers posed by continuing nuclear deployments are exacerbated by the US' and Russia's maintenance of 2,654 warheads on high alert status. This puts inhuman pressures on the leaderships of both countries who would have only a few minutes to decide whether notification of an incoming attack was accurate or a computer malfunction. The high alert status is both absurd and highly irresponsible; yet both governments opposed a resolution in the UN in 2007 to reduce the operational status of nuclear weapon systems.

Why are both states still locked in a Cold War posture? A quarter of a century ago, UN Secretary-General Javier Perez de Cuellar criticised the arrogance of the two nuclear superpowers: "By what right do they decide the fate of all humanity?" That question is still valid. Both countries continue to ignore the ruling of the International Court of Justice that negotiations for the elimination of nuclear weapons must be concluded. Such negotiations have not even begun. Nor is it likely that the other three declared nuclear weapons states, the United Kingdom, France and China, will promote or participate in comprehensive negotiations until the US and Russia drastically reduce their overwhelming numbers. There is not the slightest sign of this happening.

Worse, the US insists on deploying a missile defence system, ostensibly to counter an emerging Iranian capability, in Poland (interceptors) and the Czech Republic (radar installations). Russia has objected strenuously to the plan. Whatever the long-range repercussions of a reignited nuclear arms race, the cooperation necessary to establish confidence in negotiating further reductions has been weakened.

For its part, the North Atlantic Treaty Organisation (NATO), which still clings to a nuclear weapons doctrine, appears to be little interested in helping the international community to find security without nuclear weapons. Instead of influencing the US, their major partner, the non-nuclear states of NATO sublimate their own aspirations for a nuclear weapon free world to US dominance. Many of them admit privately that the proliferation of nuclear weapons cannot be stopped as long as major states retain their nuclear arsenals, but they say nothing. They are still resisting civil society efforts to get US tactical nuclear weapons out of five NATO countries in Europe.

At the level of existing politics, the outlook is not good, but there are new glimmers of hope on the horizon.

Much has been made of the Schultz-Kissinger-Nunn-Perry op-ed articles framing nuclear disarmament steps within the goal of abolition. Given its proponents, the Hoover Initiative puts to rest the assertion that being for the abolition of nuclear weapons is unrealistic. Since the presumptive Republican and Democratic nominees for president, John McCain and Barack Obama, have articulated progressive policies on nuclear disarmament, everyone is waiting for the next US Administration to signal at least a willingness to work cooperatively in the multilateral arena. It appears possible that an amenable US Administration will work for a successful Review Conference of the Non-Proliferation Treaty in 2010 – that is, at least not a failed conference. Because the standard for non-failure has already been set so low, it is not likely that any significant advance to a nuclear weapon free world will be made at the Review Conference. However, two events – the inauguration of a US Administration willing to respect international law and a Review Conference that keeps the NPT alive if still weak – will set the stage for some serious nuclear disarmament work to begin.

What should that work be and who should participate?

A traditional or orthodox approach would be to call for the US and Russia to negotiate a new strategic reduction treaty applying the principles of verification, transparency, and irreversibility to both delivery systems and warheads. That would be an important step. So would the entry-into-force of the Comprehensive Test Ban Treaty and successful negotiations to produce a verified Fissile Material Cut-off Treaty.

May be that is all we can get. Given the track record of both states, accomplishing this would be regarded by many as a great feat. But it will not produce the nuclear weapon free world that is not only the subject of this conference but the ardent desire of countless people around the world.

Experience has shown that a step-by-step basis does not produce nuclear disarmament. A holistic or comprehensive approach is necessary. The steps must be fitted into the vision and intention to achieve a nuclear weapon free world. As I have said, the US and Russia do not have the vision or the intention.

Left alone, both countries will do the minimum to make it appear that they are engaged in nuclear disarmament – while retaining modernised arsenals. The world will still be divided between the nuclear haves and have-nots. The unstable two-class world will go on.

In the second decade of the 21st century, renewed efforts must be made to establish a framework or convention prohibiting and eliminating all nuclear weapons. Already, a Model Nuclear Weapons Convention exists as a UN Document (A/62/650). Such an instrument can unify the non-nuclear weapon states – if they take it seriously. Like-minded governments and the advanced wing of civil society have much work to do to proclaim the Model Convention and advance its credibility.

The major role of those who believe in the necessity of a nuclear weapon free world – the number is growing – is to convince the US and Russia that the only way to stop the proliferation of nuclear weapons is to ban them entirely. The US and Russia do not believe this at the present time, but neither country is impervious to the considered views of its friends and allies. So the first thing to do to have the US and Russia exercise their responsibilities to lead the way to nuclear disarmament is for states and civil society to make this case coherently, ardently and with perseverance.

The US and Russia want everyone else to shut up while they go about trying to keep Iran and other potential nuclear weapon states from acquiring the ultimate weapon. This intimidation must be resisted. Coalitions of governments – the New Agenda Coalition is a good example – must raise their voices and connect with a strong anti-nuclear weapon public opinion that already exists within the US and Russia. Civil society movements – the Middle Powers Initiative is but one example – must work increasingly closely with like-minded governments to mount a world campaign. That campaign must culminate in a global summit for the elimination of nuclear weapons – the kind of summit called for by Kofi Annan and the Weapons of Mass Destruction Commission. The power of a sustained world movement will have a deep effect on the policies of the US and Russia.

India should be a proponent, if not the instigator, of this new world movement. I have been to India many times and I know of the longstanding Indian desire for a nuclear weapon free world. Prime Minister Indira Gandhi was the first leader to sign on to the Six-Nation Initiative in 1983. Prime Minister Rajiv Gandhi brought his Action Plan for a Nuclear Weapon Free World to the United Nations in 1988. The premise of the Action Plan is as valid today as it was 20 years ago: non-nuclear weapon states would not acquire nuclear weapons in return for a global process to eliminate all nuclear weapons.

The Western states dismissed Gandhi's plan. And, truth to tell, the changing governments of India lost heart that a country could be powerful and non-nuclear at the same time and turned India into a nuclear weapon state. Nonetheless, the vision of a nuclear weapon free world is not dead in India. Far from it. India votes regularly at the UN for active plans to reduce nuclear dangers and for a convention to prohibit not only the use of nuclear weapons under any circumstances, but also the development and stockpiling of nuclear weapons. The president of India stated on February 25, 2008:

India remains committed to universal, non-discriminatory and comprehensive nuclear disarmament as reflected in the action plan presented by the late Prime Minister Rajiv Gandhi and has called for renewed efforts for general and complete disarmament, particularly nuclear disarmament.

This very conference, "Towards a World Free of Nuclear Weapons," will end with a recapitulation of "Rajiv Gandhi's Vision of a Nuclear Weapon Free World." Clearly there is renewed desire in India to once more strike a stance to rid the world of a nuclear Armageddon. It is not too much to say that India today is at a crossroads and holds the global future of nuclear weapons in its hands. By a powerful outreach to other nations in every part of the world, India can be a catalyst to influence the US and Russia to come down from the nuclear mountain. India must become active in new efforts for global nuclear disarmament. The world will welcome India actively working with like-minded states for the advancement of human civilisation by the abolition of nuclear weapons. ∎

Need for a New Global Consensus on Non-Proliferation

☐ **SVERRE LODGAARD**

RECONFIRMING THE GRAND BARGAIN

For many years, nuclear arms control arrangements revolved around two treaties: the Anti-Ballistic Missile (ABM) Treaty and the nuclear Non-Proliferation Treaty (NPT). The ABM Treaty is gone, and the NPT is under severe pressure.

All states except four are members of the NPT. Among the four, North Korea may be ready to trade its nuclear weapons for economic benefits and normalisation with the US and the rest of the world. The remaining three – India, Pakistan and Israel – are unlikely to relinquish their nuclear weapons any time sooner than any of the five recognised nuclear weapon states (NWS). They cannot, for various reasons, become members of the NPT, but may be asked to behave "as if" they were members of the treaty, on the formula that France applied until it became a regular party in 1992.

Many NPT members stick to what is known as the Swedish example: Regardless of what the NWS do, it would not be in Sweden's interest to acquire nuclear arms.[1] In Europe, Africa and South America, many non-nuclear weapon states (NNWS) stay committed to this kind of thinking and to the NPT. In East Asia and the Middle East, where proliferation concerns are strong, most states hope that the treaty will survive. To various degrees, the P-5 also want the NPT to stay. If it unravels, there are no nuclear weapon free zones (NWFZ) to fall back upon in the Middle East and Northeast Asia, where proliferation may occur and get out of hand. International safeguards may continue on the basis of facility agreements (INFCIRC/66), but this is an incomplete substitute for NPT-type safeguards. In short, redundancies exist, but the fallback options are not attractive. For the great majority of states, breakdown of the NPT would, therefore, be a significant loss.

The best, and quite possibly the only realistic chance for a new global consensus on non-proliferation is, therefore, to reconfirm the validity of the old one, i.e. the grand bargain that was struck in the second half of the 1960s, which led to the NPT and a comprehensive international non-proliferation regime. The need to do so is urgent, for the treaty may not survive another setback at the Review Conference in 2010.

Below, ten requirements for a new global consensus are discussed. They tie into each other; the order in which they are listed is neither an order of importance nor an order of priority.

THE UNIT OF ACCOUNT

For a global consensus to be reestablished, requirement number one is that the US abandons its policy of regime change so that, once again, agreement is forged on the unit of account. In the NPT, nuclear weapons are the units of account: it is the *weapons* that should not proliferate and it is the *weapons* that should finally be eliminated. The Bush Administration shifted its attention to their possessors, of which there are good and bad ones: the president has frequently asserted that "we have to see to it that the weapons are in the right hands"/"we have to keep the worst weapons out the worst hands." The US would itself be the arbiter of the new two-tier system. This policy has to be revoked.

NEGOTIATIONS

Second, negotiations must be reinstated as a *bona fide* modality of the non-proliferation policy. As long as the leading NWS is unwilling to talk with its most ardent adversaries – to representatives of "the nexus of terrorism and weapon of mass destruction" – there can be no agreed strategy for non-proliferation and disarmament. Then, there can only be select areas of cooperation. North Korea and Iran illustrate the point. For years, the US threatened the regimes there; they have responded by pursuing their nuclear programmes; whereupon the US has used their nuclear build-up to strengthen its own case against the regimes. This way, the quest for regime change and the refusal to negotiate became proliferation drivers.

It stands to reason that under such a policy, the other side is not prepared to talk either. Knowing that the objective is to cut their throat, the adversaries have little incentive to negotiate. It was not until Berlin in January 2007 – when the US nevertheless sat down with the North Koreans – that the Korean problem took a turn for the better. Unfortunately, it happened shortly after the North Korean test, which seemed to confirm the political utility of nuclear weapons.

The P-5 plus 1 are ready to negotiate with Iran if Iran brings all fuel cycle works to a halt and accepts the Additional Protocol to the NPT safeguards agreement. This position turns what should have been the *object* of negotiations into a *precondition* for them. In Iranian eyes, it is tantamount to saying that first, you must accept defeat, and then we may sit down and talk. The key to a peaceful solution is to do in relation to Iran what has been done in relation to North Korea: first of all, the ground must be prepared for bilateral talks between the US and Iran about nuclear and other issues of common concern in the Middle East. However, the current Administration appears adamantly opposed to doing so.

PREVENTIVE MILITARY ACTION

The US National Security Strategy of 2002, updated in 2006,[2] says that the US must deter and defend against the threat of weapons of mass destruction (WMD) before it is unleashed. It will not do to wait until the targeted regimes have built their weapons and can retaliate with them. President Bush called it a strategy of preemption, in accordance with the way this expression had been used for a decade, but that use of the term blurs the important distinction between preemption and prevention. The difference is one of timing and legality. In international law, preemption – at the moment an attack is about to be set in motion – is permitted.[3] Preventive attack, against an opponent who may or may not become a real threat in the future, is not.

In recent decades, there have been four well-known cases of preventive action against actual or alleged proliferators of WMD: Osiraq, Iraq (1981), El Shifa, Sudan (1998), the occupation of Iraq (2003) and Al Kibar, Syria (2007). The results have been discouraging.

The international non-proliferation regime is rooted in international law. Illegal preventive actions undermine it. Had the intelligence about El Shifa been right – this pharmaceutical plant allegedly produced precursors for chemical weapons (VX nerve gas) and had links to bin Laden – the bombing of it could perhaps have been justified as an act of self-defence in reference to Article 51 of the UN Charter, similar to the way in which the US justified its attack on Afghanistan after 9/11, or it may have been justified as an act of reprisal.[4] It was done in response to the terror attacks on the US embassies in Kenya and Tanzania. However, eight months later, the US admitted that it had made a mistake. "Guilty till proven innocent" is, anyhow, nowhere in the legal textbook.

The actions have been marred by intelligence failures – admitted in the cases of Iraq and El Shifa; dubious diagnosis and failure to grasp the consequences of bombing in the case of Osiraq – the bombing of it accelerated the secret nuclear weapons programme that was discovered ten years later, and may even have triggered it[5]; waiting to be corroborated or falsified in the case of Al Kibar. Coming on top of the illegality of preventive actions, flawed intelligence exacerbates the negative consequences.

The political implications have been detrimental to non-proliferation endeavours. Illegal acts of war trigger strong sentiments and much enmity, especially when based on claims that are unfounded. The case of Osiraq, in particular, leaves a lesson that is pertinent to Iran and possibly also Syria. If Iranian nuclear facilities are bombed, Iran may try even harder the next time. If it was not going for nuclear weapons before being bombed, it would be more likely to do so afterwards. The Syrian case is more open to doubt. Its territory is much smaller than Iran's, and it is more vulnerable to Israeli intelligence and military capabilities than Iran.

The bombing raid at Al Kibar was the second application of the Begin Doctrine, which holds that Israel will not tolerate acquisition of nuclear weapon capabilities by its enemies (the bombing of Osiraq was the first). If the charges are correct – that Syria and North Korea were cooperating to build a nuclear reactor there – Syria had failed to declare the upcoming facility to the International Atomic Energy Agency (IAEA). Being a member of the NPT and having an INFCIRC 153 safeguards agreement in force, it should have provided design information in preparation of IAEA safeguards. Israel, which is well known for being critical of the efficiency of international safeguards, did not inform the IAEA either. In a vote of no confidence in the agency, it bombed instead. Unilateral action substituted for the way the non-proliferation regime is designed to handle such a matter, undercutting its international verification instrument.

The third requirement is, therefore, that preventive military action yields to legal, treaty-based approaches. However, military action to stop nuclear proliferation programmes is deeply rooted in US and Israeli political and military thinking. There is nothing to suggest that they will abandon such options any time soon. So the caveat is that for the NPT, the question is how best to coexist with policies which, when enacted, tend to weaken it.

Playing on the assumption that differences do not necessarily amount to incompatibilities, but can be made complementary, the tensions between regime procedures and unilateral counter-proliferation policies may be alleviated and even transformed to become mutually supportive. In certain circumstances, the international regime may stand to gain because its enforcement mechanism is commonly recognised to be weak. *(See counter-proliferation and non-proliferation, below).*

VERIFICATION

The original NPT verification standard was full-scope safeguards following all fissile materials in member states. The model agreement (INFCIRC 153) provided for more intrusive international verification than had previously been accepted by states. There were concerns in highly industrialised non-nuclear weapon states (NNWS) that industrial secrets might be revealed, but they were put to rest already by the mid-1970s.

INFCIRC 153 was deemed adequate for the purposes of the treaty. To be taken seriously, any arms control regime must have verification provisions that are considered adequate. When Saddam's secret nuclear weapon programme of the 1980s proved them inadequate, an Additional Protocol (INFCIRC 540) was elaborated that greatly improved the agency's ability to look for activities and facilities that might not have been declared. Efforts are still going on to make 153 plus 540 the new verification standard. The principle is adequacy: the operative meaning of it may have to be adjusted from time to time. INFCIRC 540 is hardly the end of the road. To oppose the Additional Protocol on the ground that it was not a part of the original bargain is, therefore, misguided. NNWS refusing to accept INFCIRC 540 as long as the NWS do not live up to their obligations should reconsider this stance. This is the fourth element to be addressed in a new search for common ground.

In 2003, the US did not allow the United Nations Monitoring, Verification and Inspection Commission (UNMOVIC) to come to a conclusion. The war cut the inspections short. The case of Iran appears similar: to keep the pressure on Tehran, it does not really want the IAEA to come to a conclusion. Some national intelligence leads have been given to the agency to follow up, but so far to little avail except for the passage of time. In essence, it is the information on the notorious Iranian laptop that ended up in the US that remains to be assessed. The US and other states claim it contains evidence of weapons intent. Iran claims it is forgery, and has not been very helpful in sorting it out.

Cooperation between the IAEA and possessors of relevant national intelligence data has been mixed. In recent years, big powers have shared some information with the agency, but there is no systematicity to it and it is not always clear for what purpose it is done. In the later phases of the United Nations Special Commission's (UNSCOM's) work, national intelligence agencies also infiltrated the Commission. Provision of national intelligence information to the IAEA for the purpose of verifying compliance with the NPT was not a part of the original bargain, and it is fraught with difficulties. But to use national intelligence to complicate and obstruct the verification process was certainly not part of the bargain either.

Reconfirmation of the basic bargain means that member states should refrain from any activity or policy that may detract from adequate verification procedures. INFCIRC 540 should be universally accepted, and those who are in a position to add to the safeguards operations of the IAEA should be encouraged to do so in good faith.

BALANCING THE RIGHTS AND OBLIGATIONS OF ARTICLE IV

Reconfirming the basics is very much about reestablishing the balance between the three pillars of the NPT: non-proliferation, disarmament and peaceful uses of nuclear energy. This is requirement number five. What has changed the original balance is first of all the failure to meet the disarmament obligation of Article VI while a number of supply-side measures have restrained the peaceful uses under Article VI, in the name of non-proliferation.

Article IV was deliberately written to avoid attempts to reinterpret the NPT in ways that would make peaceful utilisations more difficult, the only qualification being that transactions must be in conformity with Articles I and II, precluding acquisitions for weapons purposes. The right to peaceful uses includes the right to build national fuel cycle facilities, i.e. plutonium and enrichment plants.[6] The spread of such capabilities is undesirable from a proliferation point of view, but restrictions can best be promoted in conjunction with other measures to maintain the overall balance of rights and obligations. One type of trade-off that has long been discussed is assurances of supply of fuel and access to technologies in return for restrictions on national fuel cycle acquisitions.

In 1971, the Zangger Committee was established to operationalise the NPT export control provision contained in Articles III.2. In 1975, the Nuclear Suppliers Group (NSG) was established following India's nuclear explosion the year before, setting out to control the export of dual-use items. The NSG – a self-appointed group of supplier states originally referred to as the London Club, holding secret discussions in London – drew resentment right from the beginning. Critics recommended, instead, the negotiation of mutually agreed rules of international nuclear commerce between suppliers and recipients. These diverging approaches have existed ever since and they have been further accentuated recently.

Supply-side technology restraints may still be clarified, improved and consolidated. For instance, there are proposals for a division of labour between the Zangger Committee and the NSG, Zangger staying with nuclear items [especially designed or prepared (EDP) items] and the NSG with the dual-use list; for linking export control information with the IAEA verification system – current practices lack systematicity; and for negotiating a global export control convention to strengthen the standards and promote their universality and legitimacy. Clearly, there is merit to moves that can shore present arrangements up along such lines. However, the main agenda of the future should take a different tack at the problems, moving to a negotiated set of regulations that would be more legitimate, comprehensive and universal than supplier restraints can ever be, and that would provide both assurances of non-proliferation and assurances of supply and access.[7]

BALANCING THE PILLARS: THE NECESSITY TO DISARM

Such a deal has proven hard to strike, however. In view of all the attempts that have been made to date, without success, it may only be achieved in a broader framework that sees a new understanding of Article IV in conjunction with progress on Article VI. Many NNWS object to any restrictions on their Article IV rights as long as the NWS treat Article VI so lightly. In their view, the original balance of rights and obligations has shifted too much in their disfavour already.

The policies of the NWS affect the proliferation/non-proliferation process in many ways. They shape the security environment of potential proliferators. They project the utility of nuclear weapons through their strategies and doctrines: in this century, the US, Russia, France and the UK have lowered the bar of nuclear weapon use and extended the spectre of contingencies in which they might be applied. They project the status value of nuclear weapons, and they weaken the non-proliferation norm by buying selectively into the provisions of the NPT, undermining the unity of it and making it difficult to act with determination against rule breakers. To claim, as NWS sometimes do, that all of this "is inconsequential in the difficult and agonizing deliberations of a government to go nuclear is far removed from political realism."[8] Substantial progress in implementing Article VI, setting the world on a path that may lead to a nuclear weapon free world, is requirement number seven for the original bargain to be reconfirmed.

Long lists of arms control and disarmament measures have been thoroughly examined and proposed over the years, most recently by the Weapons of Mass Destruction Commission[9] and by the former US leaders behind the Hoover Plan[10]. Technically, virtually all stones may have been turned, most of them many times over. However, can arms control and disarmament policies that were conceived in the first nuclear era of the Cold War succeed under the political circumstances that underpin the second? For arms control regimes to function properly, they must reflect the distribution of power in the international system, and for many years, there has been a distinct shift of power towards Asia. China, India and Russia are increasingly self-confident and effective in protecting and promoting their national interests on the global scene.

Article VI starts with a call for "...cessation of the nuclear arms race at an early date..." This was always understood to refer to a test ban and a ban on production of fissile materials for weapons. In the bipolar world of the Cold War, it was logical to start with measures that would bring the qualitative and quantitative dimensions of the arms race to a halt and then proceed to disarmament.[11] But it did not have to be done in that order: the original bargain did not tie the members to any particular sequence. A Comprehensive Test Ban Treaty (CTBT) and a Fissile Material Cut-off Treaty (FMCT) have always been high on the list of priorities, but so have reductions in operative arsenals. The latter was never made conditional on achievement of the former. Today, the onus is on the US and Russia to take the lead and make further deep cuts in their arsenals. For China and India, doctrinal restrictions on the threat of use is another high priority.

Russia and the United States have conducted about 1,000 nuclear weapon tests each; India, 5. The advanced NWS are increasingly sophisticated in simulating nuclear explosions: the latecomers do not have the same capabilities. Will the CTBT be accepted under such circumstances? Will it have to be supplemented by NWS commitments not to introduce new types of weapons? Or should the sequencing be different, going for further substantial cuts first and trying to clinch a CTBT later? Similar questions can be asked of the FMCT. China is hedging in view of the US missile defence programme, and hesitates to proceed to an early cut-off agreement. India is in the process of building up its arsenal and is presumably unprepared as well. To Western voices advocating nuclear disarmament, Asians say, "Go ahead, we'll support you." But if the response is an arms control agenda that constrains them while leaving the usual suspects largely off the hook, initiatives will quickly go sour.

UNIVERSALITY: THE THREE-STATE QUESTION

A new global consensus must also include the three states that never joined the NPT: India, Pakistan and Israel. They are all NWS. There is no particular reason to believe that their arsenals will be more short-lived than those of the five recognised NWS. North Korea, which withdrew from the NTP in 2002, is in a category of its own. The six-party negotiations try to bring it back into the NPT as a NNWS.

After the tests in 1998, India has reworked its nuclear diplomacy. From being the world's leading protestor against "discrimination" in the nuclear order, India transformed itself into a nation ready to support the existing order. It stepped up its support for incremental arms control, endorsed the objectives of the NPT, declared its readiness to join FMCT negotiations, endorsed the NSG guidelines without being a member of the NSG, issued a moratorium on testing, supported nuclear weapon free zones elsewhere in the world, announced its willingness to seek substantive confidence-building measures (CBMs) with Pakistan, and tightened export control regulations. Later, it has supported recent US non-proliferation initiatives like the Proliferation Security Initiative (PSI).[12]

The nuclear deal with the United States rubbed some of these commitments in – the commitment to world class export controls in particular. This deal – still not clinched – also influenced India's decision to support the European resolution on Iran at the IAEA in September 2005 as well as the resolution that sent the Iranian file to the UN Security Council in February 2006. However, India is keen to preserve its independence in relation to other big powers and is likely to play an increasingly important balancing role in rapidly rising Asia.

India's orientation toward the existing order makes it easier to forge a new global consensus on non-proliferation and disarmament. To enhance and formalise its involvement in a progressive global consensus, it should be asked to behave "as if" it was a member of the NPT. The "as if" formula would oblige India not to assist others in acquiring nuclear weapons (Article I of the NPT); to abide by the rules of international nuclear commerce (the safeguards requirement of Article III.2); and to dedicate itself to nuclear disarmament (Article VI). Today, it is under no such international disarmament obligation. India has a strong tradition to uphold in this field and a legacy from Rajiv Gandhi to honour, and it maintains its support for a time-bound international convention for elimination of nuclear weapons. Therefore, this is unlikely to raise any problem.

Registration of these commitments implies recognition of India as an NWS. This makes it possible to draw it into arms control arrangements from which it is now excluded. For the time being, such a move remains controversial. The P-5 are not ready for it. However, if international law is not totally separate from international realities, sooner or later *de facto* nuclear weapon status will lead to *de jure* recognition.

The "as if" formula is applicable to Pakistan and Israel, too. Israel's declaratory policy has always been the same: never to be the first to introduce nuclear weapons into the Middle East. The ambiguity of it has served it well. To be realistic, therefore, efforts to extract non-proliferation commitments from Israel should not be predicated on a change in declaratory policy. On closer analysis, this does not pose any problem: an Israeli commitment to behave "as if" does not have to clarify whether it is an NWS or an NNWS that undertakes it. Often criticised for being recalcitrant and arrogant and for blocking arms control in a region that desperately needs it, the merit of it for Israel would be to soften that perception.

Could NPT members contemplating acquisition of nuclear arms prepare themselves under the umbrella of the treaty, give three months notice of withdrawal, and then try to soften the pressure levelled on them by offering to abide by the provisions of the NPT "as if" they were members, replicating India? This is a *bona fide* concern. There is no simple answer to it. However, unlike the three which never joined the treaty, states attempting to do this would violate an international legal commitment. Second, new procedures should be devised, anyhow, to make it harder to leave the NPT. Third, the P-5 could raise the costs of withdrawal by elaborating a set of agreed restrictions to be communicated in advance and set in motion once a member state notifies other parties and the UN Security Council that it has in mind to withdraw. India, Pakistan and Israel should be asked to participate actively in the development of such measures to ensure that proliferation becomes more, not less, difficult.

Universality was always the ambition, reiterated at all Review Conferences of the treaty. In this respect, the NPT has been more successful than any other multilateral arms control treaty. If North Korea can be brought back to the NPT as an NNWS, application of the "as if" formula to the remaining three would bring the treaty as close to universality as is practically possible. Some version of this proposition may be requirement number eight for a new global consensus to be struck.

NON-PROLIFERATION AND COUNTER-PROLIFERATION

In order for the grand bargain to be reestablished, two elements of US policy must go: the policy of regime change and the propensity to act as if the disarmament obligation of Article VI does not exist.

The US policy of regime change may be out with the next Administration. Counter-proliferation – preventive action in some form or other – is likely to stay. That goes for military as well as non-military means to eliminate nuclear installations, using conventional or nuclear weapons as appropriate, although in practice, the role of the latter may be confined to threats of use. Likewise with the notions of exceptionalism and unilateralism: the idea that there should be exceptional rules for the exceptionally strong – in high currency immediately after the occupation of Iraq – may be out, while unilateralism will stay for long. Unilateralism and counter-proliferation may change with changing administrations – there may be more or less of it – but the changes are likely to be variations of one and the same theme. These policies are structural, deriving from the superior status of the US in world affairs – from its military strength in particular – and are, therefore, likely to stay for long. The non-proliferation regime will have to coexist with these elements of US policy.

Counter-proliferation and non-proliferation policies may be incompatible. When the IAEA reports a case of non-compliance to the Security

Council, authorisation of the use of force in a counter-proliferation mode is not excluded – on questions of international peace and security, the Council is free to decide; there is nobody above it – but it is unlikely to happen. Such action is more likely to be taken unilaterally or by coalitions of the willing and be conducted in defiance of the regime. For instance, bombing of nuclear facilities in Iran would be a violation of international law, for it could not be justified as an act of self-defence in reference to Article 51 of the UN Charter – and the Security Council will not authorise it.

On the other hand, by extending credible threats of use of force, US counter-proliferation policy may be a useful supplement to the non-proliferation regime. Such threats can have a deterrent effect which the international regime has little or no prospect of providing. Deterrence does not address the root causes of proliferation, and, in that sense, does not solve the problems, but it can affect the balance of incentives in favour of staying non-nuclear. For instance, a would-be proliferator may defer a decision to go nuclear, or refrain from weaponising his programme and seek political solutions to security concerns instead.

For threats to be credible, there has to be a capability and a will to implement them. Actual use of force to destroy nuclear facilities may, however, do more harm than good. For instance, if Iran is bombed, it will leave the NPT, refuse inspections, be free to pursue a nuclear weapons programme in secret, and probably also be more motivated to do so. Bombing would set the nuclear programme back, but in the long run, it is more likely to become part of the problem than of the solution. Therefore, the relationship between counter-proliferation and non-proliferation is ambiguous and complex: counter-proliferation threats may supplement the regime in the best interests of non-proliferation, while *actual use of force* may undermine the same objective.

Requirement number nine is, therefore, to live with the differences between non-proliferation and counter-proliferation as best as we can and exploit the opportunities to turn them into mutual advantage.

Against non-state actors, collective action is more easily agreed. Security Council Resolution 1540, dealing with measures to be taken against non-state proliferators, was passed by consensus under Chapter VII of the UN Charter. For the great majority of states, violent non-state actors are "hair in the soup". Usually dubbed terrorists, they tend to be resolutely addressed, including by use of force if needed and at all possible.

LEADERSHIP

For international regimes to function properly, some important actor(s) must exercise leadership. The non-proliferation regime is no exception. This is requirement number ten. During the Cold War, the depository states took responsibility for the NPT in their special way. Recently, there has been no leadership at all.

At the Review Conferences of the treaty, disarmament has always been a sticking point. While all parties – NNWS as well as NWS – are obliged to work for nuclear disarmament, US leadership is vital. The political chain reaction started in the US, and reversal of it also has to begin in this country. The logic of the argument rests in the fact that the nuclear weapon acquisitions of the other big powers can be fully understood only in reference to that of the US, coming back to the country of origin. The realism of it lies in the fact that the US is strong on so many dimensions, and in a category of its own when it comes to high technology military strength. Therefore, Washington ought to have the confidence to seriously contemplate the idea of a world without nuclear weapons.

On three occasions recently, it has in fact proven to have that self-confidence. First, at the US-Soviet Summit meeting in Reykjavik. Second, the idea was entertained in the 1990s, on the intriguing note that in a world without nuclear weapons, the US would emerge even stronger, for nobody is in the neighbourhood of matching US conventional strength. Third, it came to the fore again in connection with the 20th anniversary of Reykjavik in the form of the Hoover Plan. Three camps are likely to shape the US debate about it: the Shultz-Nunn-Perry-Kissinger camp which holds out the vision of total disarmament; a "the vision is right but unrealistic" camp that wants smaller inventories; and a "this is a very bad idea" camp that feels and fears it is a return to arms control.[13]

In the US House of Representatives, a bipartisan resolution has been introduced calling for a reduction of deployed US and Russian nuclear weapons to no more than 1,000, and a total (including reserves) of no more than 3,000.[14] This would be a significant move, but to bring the other NWS into multilateral disarmament negotiations, the arsenals must be reduced to the level of hundreds. The Hoover Plan emphasises the relationship between vision and measures, and the further down one goes, the tighter that relationship becomes. A world free of nuclear weapons should be a world where not only the weapons, but also all dedicated nuclear weapon infrastructures are gone. If not, one would end up with a number of *threshold states*, and that is an unstable and untenable proposition. At the moment, all NWS are modernising and enhancing their nuclear capabilities. They are hedging their bets, emphasising that the stronger their capabilities to regenerate nuclear forces, the further down they can go. At advanced stages of disarmament, this would have to change.

Should the US be willing to take the lead, the most vulnerable NWS will be the hardest to convince about the merits of nuclear weapon freedom. Russia says that the larger role accorded to nuclear weapons in its 2000 military doctrine is temporary, yet for some years to come, it may not be prepared for a radical turn of trends toward zero. Israel, where the nuclear arsenal is the ultimate national insurance premium in a volatile Middle East, may be the hardest case of all. Its declaratory policy of opacity blocks any participation in nuclear arms control and disarmament discussions. In these cases, it is not enough to eliminate the nuclear arsenals of those that are higher up the chain. Conventional force adjustments and credible security assurances in the form of peace agreements are of the essence. ■

NOTES

1. The only reservation was in reference to possible long-term developments making nuclear weapons standard equipment of ordinary states. Jan Prawitz, *From Nuclear Option to Non-Nuclear Promotion: The Sweden Case*, Research Report No.20, The Swedish Institute of International Affairs, 1995.

2. The National Security Strategy of the United States of America, the White House, March 16, 2006.

3. Such as when military units begin to move in preparation of an attack. Terms like "about to attack" and "seeing the attack coming" leave much to interpretation, however. The downside of it is that preemption may mean inadvertent escalation to war. For instance, during the Cold War, the offensive capabilities along the dividing line in Central Europe were commonly recognised to outweigh the defensive capabilities on the other side. Therefore, the temptation to preempt would have been strong had the one or the other side begun to believe that an attack was coming. Preemption of a terrorist attack is quite another matter: in this case, there may be no immediate warning signals, leading into an argument for preventive action against terrorists.

4. International law recognises a limited right of reprisal in addition to the right of self-defence. It legitimises certain actions that would otherwise be unlawful, in response to an illegal act by, or under the aegis of, another state. Reprisals, too, are subject to the general requirements of necessity and proportionality. Non-coercive means should be considered before a reprisal is carried out. There is a substantial body of legal opinion holding that permissible reprisals are limited to measures other than armed force.

5. Osiraq was built to run on highly enriched uranium (HEU) and was a poor plutonium source. Had a natural uranium blanket been put around the core, some plutonium could have been produced, but without a major reconstruction of the reactor, it would not have been enough for weapon production. The HEU that had

been stored at the reactor site was not hit by the bombs and stayed there until shortly before the 1991 War, when Saddam ordered it removed in a futile, last-minute effort to get a bomb. Had the Iraqis begun to adapt the reactor to make it a better source of plutonium, or removed the HEU it was designed to use, the alarm bell would have rung immediately, because the reactor complex was under IAEA safeguards. Dan Reiter, "Preventive Attacks Against Nuclear, Biological and Chemical Weapons Programs – The Track Record" in William W. Keller & Gordon R. Mitchell, eds., *Hitting First: Preventive Force in US Security Strategy*, (University of Pittsburgh Press, 2006).

6. The right to enrich is spelt out convincingly and in all necessary detail in Steven E. Miller, "Proliferation Gamesmanship: Iran and the Politics of Nuclear Confrontation," paper prepared for the symposium on *A Nuclear Iran: The Legal Implications of a Preemptive National Security Strategy*, Syracuse University School of Law, Syracuse, New York, October 26-27, 2006.

7. Tariq Rauf, "International Export Controls and Multilateral Nuclear Arrangements", in Morten Bremer Mærli & Sverre Lodgaard, eds., *Nuclear Proliferation and International Security*, (UK Routledge, 2007). In a much noted article published in the autumn of 2003, Mohamed El Baradei proposed a sequence of measures which has later been refined and followed up upon, and which represents a constructive agenda for further inquiry and action. The elements are: assurances of supply (of fuel) and access (to equipment and technology); a moratorium on new national fuel cycle facilities; restricting enrichment and reprocessing operations to facilities under multinational control; investments in new proliferation-resistant nuclear energy systems; and multinational approaches to the management and disposal of spent fuel and radioactive waste (El Baradei 2003). The High Level Panel on Threats, Challenges and Change supported many of these measures (United Nations 2004). So did Kofi Annan's report *In Larger Freedom* (2005).

8. Harald Muller, "The Future of Nuclear Weapons in an Interdependent World", *The Washington Quarterly*, Spring 2008.

9. *Weapons of Terror: Freeing the World of Nuclear, Biological and Chemical Weapons*, Report by the Weapons of Mass Destruction Commission (under the chairmanship of Hans Blix), June, 2006.

10. Shultz, Perry, Kissinger and Nunn, "A World Free of Nuclear Weapons," *Wall Street Journal*, January 4, 2007, and Shultz, Perry, Kissinger and Nunn, "Toward a Nuclear-Free World," *Wall Street Journal*, January 15, 2008.

11. During the NPT negotiations, the analogy was made to a motor car: to put it in reverse gear, you must first stop it.

12. Raja Mohan, "India's Nuclear Exceptionalism," in Morten Mærli & Lodgaard, eds., n. 7.

13. Communication from John Hamre, director of the Centre for Strategic and International Studies (CSIS) and chairman of Pentagon's Defence Advisory Board.

14. Introduced by Representatives Dan Lungren (Republican) and Jim McGovern (Democrat) in March 2008.

Managing the Nuclear Industry in a World Without Nuclear Weapons

☐ **GEORGE PERKOVICH** AND **JAMES ACTON**

INTRODUCTION: KEEPING THE WORLD SAFE IN THE NUCLEAR RENAISSANCE

Calls for nuclear disarmament are intensifying just as nuclear energy is expected to expand greatly worldwide. Much greater tension exists between these two objectives – nuclear disarmament and expansion of nuclear energy – than has been publicly discussed. Following the failure of the Baruch Plan in late 1940s, the total elimination of nuclear arsenals almost completely fell off the practical international agenda until after the Cold War ended.[1] It did briefly resurface with Gorbachev's 1986 call for nuclear abolition, the Reykjavik Summit in October that year, and Rajiv Gandhi's 1988 speech to the UN – but the accident at Chernobyl in April 1986 had put an end to any hopes of a significant expansion in a nuclear industry that had already been in grave difficulty for a few years. The nuclear Non-Proliferation Treaty (NPT) Extension Conference of 1995 and the Review Conference of 2000 put disarmament back on the agenda, albeit precariously. But, at that time, the nuclear industry was still in the doldrums.

In other words, since the 1940s, when the Baruch Plan and its predecessor, the Acheson-Lilienthal Plan, sought to prevent proliferation by taking control of the fuel cycle from the hands of any individual state, the challenges of nuclear disarmament have not been addressed *together with* the management of an expansion in nuclear energy.

The potential expansion of nuclear energy globally carries proliferation risks if there are not new, reliably-enforced rules to manage it. But key non-nuclear weapon states already express deep reluctance to consider new rules necessary to make nuclear energy secure if the nuclear-weapon states do not undertake a yet-to-be-defined plan for nuclear disarmament.[2] However, the nuclear-armed states would not agree to eliminate their nuclear arsenals if they were not confident that proliferation will be prevented through the enforcement of stronger non-proliferation rules. This is a circular problem to whose resolution the present paper seeks to contribute.

This problem is exacerbated by a second inequality: one between states that possess enrichment or reprocessing capabilities and those that do not. If there is to be a significant expansion of nuclear energy, global capacity to manufacture nuclear fuel will have to be increased. Many international leaders recognise that the spread of fuel cycle facilities to non-nuclear weapon states poses an inherent proliferation risk.[3] States that possess such facilities for civilian purposes could use them, or the associated knowhow, to produce fuel for weapons. With demand for nuclear fuel projected to rise dramatically, several states, including Argentina, Brazil, Canada, Iran and South Africa have expressed an interest in developing enrichment programmes or have already begun to do so. Yet, these states object (more or less strenuously) to rules precluding the spread of dual-use fuel cycle capabilities, partly because they would further entrench the existing inequality between fuel cycle 'suppliers' and 'recipients'. And if more non-nuclear weapon states, such as those mentioned directly above, develop enrichment capabilities before new rules are enacted, resistance to such rules will only be intensified further, especially in the Middle East and Asia.

Proposals to resolve this central dilemma are currently being developed. A number of states, the World Nuclear Association, the International Atomic Energy Agency (IAEA) and the Nuclear Threat Initiative have floated various mechanisms for assuring fuel supply to states.[4] All of them are put forward in the hope that states will choose to eschew new national facilities to enrich uranium or separate plutonium (and, indeed, this is an explicit condition of some of the guarantees). Optimally, many of these actors would prefer that a global rule be created to ban the spread of fuel production to states that do not already possess such facilities. However, because key non-nuclear weapon states, including Brazil, Egypt, Iran, and South Africa reject any binding rules, voluntary restraints seem to be the most that the political traffic will bear in the near term. But voluntary agreements whereby a state agrees to forego indigenous fuel production in return for international nuclear cooperation on a case-by-case basis provide less-than-robust confidence that proliferation will be avoided over time. If states that aspire newly to develop major peaceful nuclear programmes will not accept binding rules to forego enrichment and reprocessing, are there other measures they would endorse to improve confidence that nuclear proliferation will not occur, recognising as discussed below that current safeguards are potentially inadequate in a disarming world?

In a sense, current nuclear suppliers – many of whom now live under nuclear deterrent umbrellas – and aspiring sellers and buyers are talking past each other. The former are looking for strong bulwarks against future proliferation while the latter want to keep their options open. At a minimum, most buyers do not want to surrender their "nuclear rights" and at a maximum, want to hedge against future insecurity. What's been absent is direct bargaining wherein suppliers and buyers clearly articulate their interests and the potential trade-offs they are prepared to negotiate. Developing country non-nuclear weapon states have tended to refrain from creative give-and-take in addressing the global fuel cycle challenge. Perhaps this non-specificity reflects a natural tendency of the weaker party in a negotiation to hear what the stronger has to offer, or the relative lack of nuclear expertise. Whatever the cause, the reticence of future nuclear 'buyers' leaves many questions unanswered about the future of nuclear energy globally and the evolution of the non-proliferation regime. If this web of issues is not disentangled, the latter steps toward nuclear disarmament likely will not be taken, though this need not preclude many earlier steps in this direction.

Another major potential tension between the growth of nuclear energy and the elimination of all nuclear arsenals centres on global shortages in capacity to produce nuclear reactor components. For the next ten years, the world's nuclear industry probably can build no more than 6-7 reactors per year.[5] Suppliers for the first time will likely have more opportunities than they can service, and will pursue those with the least risk of disruption, such as political turmoil, liability or payment disputes. Those opportunities will tend to be in China, the US, South Korea and Europe and possibly India. Established vendors will be less interested in non-nuclear weapon states in the developing world, especially in the less politically stable ones. If developing countries seeking nuclear cooperation are rebuffed, and feel that their right under Article IV of the NPT to assistance in developing nuclear technology for peaceful purposes is unrealised, they could then become still further alienated from the non-proliferation regime. As a result, key non-nuclear weapon states could become less supportive than they are today of rules to manage the nuclear industry, including the fuel cycle. This friction could make nuclear-armed states less willing to disarm.

While it is too early to know whether supply constraints will engender a backlash against efforts to strengthen the non-proliferation regime, representatives from states that are able to supply nuclear technology and expertise and those from states newly aspiring to develop nuclear industries should address these issues candidly. The IAEA is now constructively engaging relevant parties on these matters. Such discussions could be broadened to include civil society in developing countries and leading potential vendors.

Not only is there a shortage in the global capacity to manufacture reactors, there is also a skills shortage. Even without a nuclear renaissance, finding enough inspectors to implement all the verification measures necessary to facilitate disarmament would be a challenge. If the inspectorate has to compete against an expanding nuclear industry for personnel, the problem would be exacerbated yet further. Clearly, if disarmament is to be effected alongside a global expansion of nuclear energy, considerable investment in training would be required.

Today, nuclear weapon states, and their allies, are willing to tolerate weaknesses in safeguards in non-nuclear weapon states partly because they own nuclear weapons which ameliorates the consequences of potential proliferation. Whether or not a rational cost-benefit analysis would conclude that in today's world safeguards should be strengthened, they would almost certainly need to be improved significantly if disarmament were to be taken seriously. This is true regardless of whether attempts to prevent the spread of fuel cycle technology are successful.

The challenge is complicated by the long time-scales involved. Not only would disarmament require a significant length of time to effect, but vigilance would be required long afterwards to prevent rearmament. In recent years, the difficulty of getting centrifuge technology to work has been an important barrier against unsafeguarded fissile material production – at least, that seems to be a lesson from Libya, and until recently, Iran.[6] Iraq suffered similar, if not such acute, difficulties.[7] This barrier may, however, be gradually eroded as more states acquire advanced industrial bases. In addition, it is also possible that over the long-term, other enrichment technologies that can be concealed more easily than the gas centrifuge will come into play. In this regard, laser enrichment – if it can be made to work on a commercial scale, thereby, increasing the risk of technological diffusion – is of particular concern.[8]

There needs to be additional research and debate on how the expansion of nuclear energy can be made compatible with progress toward eliminating all nuclear arsenals. It needs to involve experts beyond the nuclear industry and nuclear weapon establishments. Non-nuclear weapon states must be brought more fully into the process, with an understanding that emphasis should be on the practical issues at hand rather than more global inequities which would be better addressed in other forums. If governments lag in sponsoring such interactions, non-governmental actors should fill the void. As a contribution to this debate, our forthcoming *Adelphi Paper* summarises options for strengthening control of the nuclear industry. They range from incremental improvements in existing safeguards to the much more radical option of eliminating the most proliferation sensitive activities. We discuss here only one of the more prominently mentioned approaches: multinational or international ownership.

MULTINATIONAL OR INTERNATIONAL OWNERSHIP

One alternative to traditional safeguards on nuclear facilities owned by states is for the fuel cycle to be multinationalised (that is, facilities are owned and operated by a coalition of states) or even internationalised (ownership and operation by an international body) as envisioned by Acheson-Lilienthal. Indeed, some have gone so far as to argue that it is impossible to move to a nuclear weapon free world without first placing all enrichment and reprocessing facilities (and possibly all nuclear materials as well) under multinational or international control (and still under IAEA safeguards).[9] Would India, for instance, be willing to dismantle its last nuclear warhead if the Khan Research Laboratories in Pakistan were still enriching uranium under exclusive Pakistani control, even under the watchful eye of IAEA inspectors? Even if it were (and that is far from clear), it would almost certainly be on the condition that India continued plutonium production under exclusively Indian control, in which case the existence of nationally controlled fuel cycle capabilities could test the stability of Indo-Pakistani relations. In another part of the world, if Japan and China were continuing national fuel cycle activities, for peaceful purposes and under safeguards, would Asian states seek similar national fuel cycle capabilities? The chances of them doing so might be lessened if greater non-nuclear extended deterrence from the US were offered, but that could raise other alarms.

Moving beyond nationally-owned fuel cycle facilities could potentially be a key step toward disarmament – and it is a concept states should take seriously and discuss. Specifying and implementing the procedures to make this transition would be terribly complicated and politically challenging.[10] There is no analogy of an international monopoly controlling a key facet of a major modern industry. Today, the idea meets firm resistance from almost every state and enterprise now producing fissile materials, especially the states with nuclear weapons. Nevertheless, there are two precedents for multinational enrichment companies in the form of the Uranium Enrichment Company (Urenco) and EURODIF. Indeed, the former, being a private company, is potentially of particular significance.

In any case, multinationalisation or internationalisation of the fuel cycle would not completely assuage proliferation concerns. The problem of clandestine fuel cycle facilities would remain. Even though multinational or international ownership could help to alleviate this problem to some extent by restricting or fragmenting knowledge (depending on the exact model), some workers would still learn proliferation-sensitive information. Indeed, the infamous A.Q. Khan network grew out of the Urenco consortium insofar as Khan was employed by one of its contractors and stole blueprints, components and valuable procurement information from it. At that time, Khan was only a junior scientist employed in a fairly limited role. For a company to run effectively, there must be senior managers in a multinational fuel cycle consortium with a good knowledge of the entire process, and they could put that expertise to nefarious uses.

Moreover, there is an important difference between ownership and control. Shared or international ownership may make it more embarrassing for a state to steal nuclear material from a facility on its territory, and increase the consequences of it doing so, but that would not necessarily stop it. States would need to assess the risks of a host government 'sending in the troops', physically taking control of an enrichment or reprocessing plant and using it to produce fissile material for weapons purposes.

In theory, this problem might be alleviated by locating fuel cycle facilities on the territory of 'completely trustworthy' states. In practice, public pressure may make such states reluctant to play host, especially if the facility in question were a multinational reprocessing facilities responsible for dealing with imported nuclear waste. Besides, for the sake of equity, multinational fuel cycle facilities would likely need to be hosted by a number of states in different regions. The chance of the international community agreeing upon which states are 'completely trustworthy' appears slim, at least from today's perspective.

There are many different possible models for multinational or international control of the fuel cycle. Key issues include which facilities should be included (enrichment and reprocessing plants or all nuclear facilities) and how these facilities would be owned and operated. States, the nuclear industry and wider society should start considering different models to decide which are best at assuaging proliferation concerns, which are the most politically feasible and which make the most economic sense. These considerations should help shape the important debate in today's world about how to provide a guarantee of supply to states that do not have the capability to manufacture their own nuclear fuel. It is vital that such discussions include potential suppliers and consumers.

Perhaps the most that can be said now about these important and difficult issues is that a non-discriminatory framework would be necessary. Non-nuclear weapon states are unlikely to agree to new rules or arrangements for limiting access to fuel-cycle capabilities if all states do not have to play by the same rules. Genuine commitment and movement toward nuclear disarmament could satisfy this demand for equity. Key non-nuclear weapon states make this point today: they will reject discriminatory approaches to fuel cycle management, therefore, the choices are that either all states have the same rights to enrich uranium and separate plutonium in the same way as the nuclear weapon states do, albeit under safeguards, or, all states get rid of their nuclear weapons and adopt a uniform, internationalised approach to fuel production. This challenge will be among the most crucial and vexing that states will face in the nuclear realm whether or not nuclear abolition becomes a priority. ∎

NOTES

1. One exception was the McCloy-Zorin Accords of September 1961 which set out a series of "principles as the basis for future multilateral negotiations on disarmament." For a summary of the various "abolitionist waves," see Michael Krepon, "Ban the Bomb. Really," *The American Interest*, vol. 3, no. 3, January/February 2008, pp. 88-93, http://www.the-american-interest.com/ai2/article.cfm?Id=371&MId=17 (subscription only).

2. James M. Acton, "Strengthening Safeguards and Nuclear Disarmament: Is There a Connection?" *The Nonproliferation Review*, vol. 14, no. 3, November 2007, pp. 525-535.

3. See, for instance, "Address to the Nuclear Non-Proliferation Treaty Review Conference," speech by UN Secretary General Kofi Annan, New York, May 2, 2005, http://www.un.org/events/npt2005/statements/npt02sg.

4. Tariq Rauf and Zoryana Vovchok, "Fuel for Thought," *IAEA Bulletin*, vol. 49, no. 2, March 2008, pp. 33-37, to appear at http://www.iaea.org/Publications/Magazines/Bulletin/.

5. Mark Hibbs and Ann MacLachlan, "Vendors' Relative Risk Rising in New Nuclear Power Markets," *Nucleonics Week*, vol. 48, no. 3, January 18, 2007. See also Sharon Squassoni, "The Realities of Nuclear Expansion," pp. 7-10.

6. Iran's centrifuge programme was initiated in 1985 and was still fairly embryonic when it was revealed in 2002. Although there may be a number of reasons for this slow pace of development, one must surely be the difficulty of centrifuge technology. IAEA, "Implementation of the NPT Safeguards Agreement in the Islamic Republic of Iran," GOV/2004/83, November 15, 2004, para. 23, http://www.iaea.org/Publications/Documents/Board/2004/gov2004-83.pdf. For a discussion of Libya's centrifuge programme see Wyn Q. Bowen, *Libya and Nuclear Proliferation*, chapter 2.

7. Mahdi Obeidi and Kurt Pitzer, *The Bomb in My Garden* (Hoboken, NJ: John Wiley & Sons, 2004), chapters 3-6.

8. Allan S. Krass, Peter Boskma, Boelie Elzen and Wim A. Smit, *Uranium Enrichment and Nuclear Weapon Proliferation* (London: Taylor & Francis for SIPRI, 1983), chapter 2, http://www.sipri.org/contents/publications/Krass83/Krass83.pdf; Jack Boureston and Charles D. Ferguson, "Laser Enrichment: Separation Anxiety," *Bulletin of the Atomic Scientists*, vol. 61, no. 2, March/April 2005, pp. 14-18,

http://thebulletin.metapress.com/content/p3v65u514051j882/fulltext.pdf (since this article was published General Electric has acquired the rights to the SILEX process).

9. For instance, Marvin Miller and Jack Ruina wrote, "The only real control of breakout in a NWFW is strict international control of all facilities for the production of fissionable materials that could be used in nuclear weapons. We do not address the political feasibility of such controls." Marvin Miller and Jack Ruina, "The Breakout Problem" in Jack Steinberger, Bhalchandra Udgaonkar and Joseph Rotblat, eds., *A Nuclear-Weapon-Free World: Desirable? Feasible?*, p. 101.

10. IAEA, "Multilateral Approaches to the Nuclear Fuel Cycle," INFCIRC/640, February 22, 2005, http://www.iaea.org/Publications/Documents/Infcircs/2005/infcirc640.pdf.

Role of CBMs in Promoting Nuclear Disarmament

☐ LI CHANG-HE

OVER half a century, it has been a common concern and strong aspiration of the international community to put an end to the nuclear arms race, reduce nuclear arsenals and prevent unprecedented holocaust to mankind with the use of these most devastating weapons of mass destruction, either in a nuclear war or in an accidental event. The last century saw several waves of thought worldwide calling for complete nuclear disarmament. In the past couple of years, it seems that a new wave of campaigns is surging for eliminating nuclear weapons , with well-known proposals put forward in the Report of Weapons of Mass Destruction Commission (WMDC) headed by Dr. Hans Blix, former director-general of the International Atomic Energy Agency (IAEA), and in the essays and articles published by Robert McNamara, former US secretary of defence, George Shultz, Henry Kissinger, former US secretaries of state and some other former US senior officials as well as experienced arms control experts.

Among other things, practical, workable and appropriate confidence-building measures (CBMs) can play a measured positive role in promoting nuclear disarmament.

In this connection, as the most important and primary CBMs, it is imperative to build and enhance strategic trust in political and security areas between the parties concerned.

Nuclear disarmament has a close bearing on national and global security. In the course of globalisation no country can ensure its security on its own. Countries need to rely on cooperating with each other to effectively address various security issues and meet common challenges.

A sound international security environment lends an impetus to, and guarantees, nuclear disarmament; and progress in nuclear disarmament will, in turn, help lay a solid foundation for international security. In view of this, all states should work together to create an international environment of peace, stability and common security by adhering to the purposes and principles of the UN Charter, following the principles of peaceful coexistence, mutual respect for sovereignty and territorial integrity, mutual non-aggression, non-interference in each other's internal affairs, equality and mutual benefit, as well as the international law and other universally acknowledged norms of international relations. Countries should abandon the "Cold War mentality", reduce the role of nuclear weapons in world politics and national security policy and cooperate in constructing a harmonious world featuring mutual trust, mutual benefit, equality and cooperation, so as to eliminate fundamentally the motivation for maintaining and acquiring nuclear weapons, and create necessary conditions for realising the goal of nuclear disarmament and non-proliferation.

Secondly, if all nuclear weapon states are committed to not being the first to use nuclear weapons (no first use) it would constitute a confidence-building measure of great significance. Though it is a declaratory step, nevertheless it will greatly restrain the value of possessing nuclear weapons and dampen the desire to pursue them. Doing so would facilitate nuclear disarmament towards their complete prohibition and thorough destruction, following the example of the total ban of chemical weapons, a whole category of WMD. It is known to all that the 1925 Geneva Protocol banning the use of chemical weapons was actually a no first use international treaty. On this basis, the international community made untiring concerted efforts for 71 years more until the final conclusion of an international treaty that prohibits and abolishes chemical weapons completely. When all nuclear weapon states renounce the policy of nuclear deterrence based on first-use, it would give great impetus to the nuclear disarmament process.

Thirdly, in the implementation of nuclear arms reduction agreements, it is essential for nuclear weapon states to abide by the spirit and principles of maintaining global strategic balance and stability without impairing the security for all other states. Nuclear disarmament should be a justifiable and reasonable process of gradual reduction in a downward spiral. To this end,the two states possessing the largest nuclear arsenals bear special and primary responsibilities and should take the lead to reduce their nuclear arsenals significantly in a verifiable and irreversible manner so as to create conditions for other nuclear weapon states to take part in the process of nuclear arms reduction in due course and achieve the ultimate goal of complete nuclear disarmament. It is hoped that these two largest nuclear weapon states will continue to drastically reduce their nuclear armaments.

Meanwhile, it should be especially noted here that the establishment of a global network of missile defence systems, including its deployment in certain localities and relevant activities in partnership, is not conducive to global balance and stability and would hinder international nuclear arms control efforts, weakening confidence-building and trust among countries. In addition, once outer space is weaponised, it would inevitably impair and undermine efforts for nuclear arms control, disarmament and non-proliferation. Therefore, the abandonment of the global missile defence programmes and the commitment not to place weapons in outer space by all states would also be a confidence-building measure that has an important bearing on the nuclear disarmament process.

Fourthly, in order to maintain and enhance strategic trust and restraint between the nuclear weapon states in the course of nuclear arms control and reduction, they should all diminish the role of nuclear weapons in their national security policies by, for instance, no longer conducting R&D for any new type of nuclear weapons, particularly low-yield nuclear weapons useable at a lower threshold.

Other confidence-building measures for nuclear arms reduction are, *inter alia*:

- Signing and ratifying the Comprehensive Nuclear Test-Ban Treaty (CTBT) for early entry into force of the treaty, while the nuclear weapon states remain committed to the moratorium on nuclear testing.
- Starting negotiations on banning the production of weapon-grade fissile materials with a view to concluding a Fissile Material Cut-off Treaty (FMCT).
- Commitment by nuclear weapon states not to target their nuclear weapons at other countries,nor to list any countries as targets of a nuclear strike (de-targeting).
- Removing of nuclear weapons from hair-trigger alert status (de-alerting) and as a step further, dismantling warheads from nuclear-tipped missiles (de-mating) so as to increase early warning time and reduce the danger of accidental, unauthorised or mistaken use of nuclear weapons.
- Withdrawing and bringing back home the nuclear weapons deployed by nuclear weapon states outside their own territories.
- Abandoning the policy and practice of a nuclear umbrella.
- Enhancing transparency and mutual understanding between the parties concerned without jeopardising national security interests through dialogue, talks, information sharing, exchange of visits, and hot line communication at various levels.

CONCLUDING REMARKS

The CBM is by no means a stereotyped and static concept. In fact, once the necessary conditions are ripe for a certain CBM, for instance, when all major states have taken the same measure, they can move it further and gradually turn it into a legally binding rule. If the above-mentioned CBMs can be implemented and elevated to a legal level, it would facilitate substantial and significant progress in nuclear disarmament.

Meanwhile, it should also be noted that as nuclear arms reduction and disarmament concerns the vital national security interests of the countries concerned, it is a highly sensitive and complex issue. Consequently, it is only natural that CBMs in this field are also very sensitive and complex. Therefore, in exploring and developing CBMs that may be acceptable to all relevant parties, there is a need to take into consideration how much sensitivity they can endure and the degree of comfort of these parties in order to seek as much common ground as possible, taking an objective, earnest and down-to-earth approach in a step-by-step and incremental manner. The next couple of years might be a crucial period for international nuclear arms control and disarmament. The new president in the United States would in due course review the US policy on arms control and disarmament; the foreign policy of Russia under President Dmitry Medvedev is also a matter of universal attention; and the upcoming 8th Non-Proliferation Treaty (NPT) Review Conference is scheduled to be held in 2010. The world has to wait and see whether there will be a favourable turn for the better from the decade-long logjam so far in the course of nuclear disarmament and non-proliferation. ■

Reducing Reliance on Nuclear Weapons

☐ **IVAN SAFRANCHUK**

RELIANCE on nuclear weapons is not easy to discuss, because the nuclear states' reasoning in favour of nuclear weapons is often unclear and confusing. The outstanding example is the case of US tactical nuclear weapons in Europe. Their numbers have been significantly downsized after the end of the Cold War, but still over hundred remain in Europe. If asked, US and European politicians give explanations far from military utility or political rationality. It is somewhat like "just in case we need them." French nuclear strategy is another example of possession of nuclear weapons becoming a value in itself.

Still one common point may be found for the nuclear strategies of nearly all nuclear powers, including those not identified in the Non-Proliferation Treaty (NPT).

Nearly all the nuclear powers, with only one exception to be discussed below, regard their nuclear weapons as part of the defence posture. There are two basic strategies – to win a war and to deny victory to an enemy. Most nuclear powers are not going to win wars with nuclear weapons. They maintain nuclear weapons as weapons of last resort, to refer to them when and if the fundamentals of their existence are at stake. Not in any other case.

It is important to understand that most of the nuclear powers regard their nuclear weapons as weapons of necessity (not of choice) and of last resort.

The only exception from the general logic is represented by the US. This country has been constantly looking for opportunities to make nuclear weapons more usable. While all others are interested, within their defensive approaches, in keeping the threshold of use of nuclear weapons as high as possible, in the US there was pressure to decrease the nuclear threshold and make nuclear weapons more usable on the battlefield. Technical efforts have been invested into achieving this goal. Hopefully, the new Administration will hear the critics of this approach and abandon the efforts.

To sum up, all the nuclear powers, except the US, keep nuclear weapons as weapons of necessity and last resort, and do not take any steps to lower the nuclear threshold and make them more usable. The US has tried to make nuclear weapons more usable, but is likely to retreat on this.

Proliferators or suspected proliferators also develop nuclear weapons within the defensive rather than offensive logic.

Lack of security in the Middle East drove Israel to acquire nuclear weapons, not because it intends to use them in offence or preemption, but as a defensive means. India and Pakistan also acquired nuclear weapons to improve their defence, to make it absolute, rather than to use first in a war of choice. Iran and North Korea are neither driven by offensive motivations in their presumed programmes.

I do not value the high so-called status motivations in favour of nuclear weapons. They do exist. But in most cases, they are at least secondary to hard security considerations.

For the majority of nuclear powers and proliferators, reliance on nuclear weapons is the reaction to insecurity, to the increasing role of force and violence in world politics. It is absolutely rational to think that if you have nuclear weapons, nobody will start a war of choice against you.

There is the theory of "democratic peace," stating that democracies do not fight each other. "Nuclear peace" has the same mathematics behind it. Nuclear powers do not fight each other, at least not directly. The reality may be that nuclear conflict is possible between nuclear states, but the prevailing perception is that a nuclear power will not find itself under aggression.

Can a world of violence and force be free of nuclear weapons? My answer is no.

If so, then on the way toward a world free of nuclear weapons, the major obstacle is the current world disorder, where a country may be attacked just to find in the end that the cause for the attack was flawed, if not manipulated on purpose. In a world where order and justice are next to non-existent, where a word has be enforced with power to be heard and taken into consideration, and where international law is neglected, nuclear weapons cannot but be attractive.

So the core of the efforts to build a world free of nuclear weapons should be to build a world ruled by international law, not by force.

Here, I think, is the key difference between those who approach the nuclear free world option from moral and realistic perspectives. For those, who come from the moral side, a world free of nuclear weapons is a more secure and more just world, where international law and/or human principles prevail. For those, who come from the realistic perspective, a world free of nuclear weapons is not necessarily a world ruled by international law. Why do prominent US realists send signals of readiness to go beyond traditional arms control, up to the vision of a nuclear free world? This is the result of the long debated argument that spread of nuclear weapons is the ultimate limitation for US capability to project power globally. With the current conventional superiority, in a world free of nuclear weapons, the US would enjoy much more freedom and opportunity to use force than in the current nuclear world.

The paradox is that most nuclear powers would probably feel less secure and more vulnerable if the world, as it is now, were to suddenly become free of nuclear weapons. And those who would feel more free and less constrained after being disengaged from nuclear deterrence would not necessarily use this for the interests of peace.

To make possible a world free of nuclear weapons, many things in the world would have to be fixed. There has to be a real programme of adjusting conventional imbalances (to persuade those who bridge their conventional weakness with the ultimate power – nuclear – of defence, to give up the latter), improving regional security systems in the Middle East and South Asia, and making real progress toward the rule of international law.

Symbolic measures are needed. There should be more looking into the cases of countries that took the decision to give up their nuclear arsenals. There is a range of samples already: the South African experience; the stories of Ukraine, Belarus and Kazakhstan, that smoothly (except in the Ukrainian case) gave up nuclear weapons, which accidentally fell into their hands after the collapse of the Soviet Union; Libya gave up its nuclear programme. But even more emphasis should be put on the research and promotion of the earlier examples. Dozens of countries went through nuclear temptation and chose to be non-nuclear, despite technical feasibility in many cases. The experience of abandoning nuclear weapons does exist, as well as the wide experience of choosing not to be nuclear.

A strong symbolic measure could be represented by withdrawal from nuclear security guarantees by those who enjoy them. For many countries that enjoy nuclear guarantees of the major great powers, it may be easier to withdraw from such arrangements than for the nuclear powers themselves to abandon nuclear weapons. This would be a strong sample of decreasing reliance on nuclear weapons.

Propaganda about the nuclear dangers can help achieve some goals and raise public opinion and among civil groups. However, this is far from being the remedy. Most nuclear states are not fans of nukes, they treat them as weapons of necessity and last hope, while admitting their dangers.

Whatever positive examples the symbolic measures may present, still the core of the programme toward a world free of nuclear weapons should be to focus on hard security and rule of international law. Choosing the correct way toward a world free of nuclear weapons is crucially important. The approach to focus primarily or even exclusively on delegitimisation of nuclear weapons and treat nuclear weapons separately from other security issues – this is the way to nowhere. To concentrate on propaganda about nuclear dangers, raise public concerns, develop a convention and then mount pressure on nuclear states to join it – this is an exclusive, narrow and antagonistic approach.

Actually Rajiv Gandhi was not speaking about such approach. This great politician of the world envisioned a comprehensive approach toward a world free of nuclear weapons. This whole speech, delivered at the UN General Assembly on June 9, 1989, was about a wide approach. Below are three citations exactly on this:

> ... changes are required in doctrines, policies and institutions to sustain a world free of nuclear weapons. Negotiations should be undertaken to establish a Comprehensive Global Security System under the aegis of the United Nations.

> The plan for radical and comprehensive disarmament must be pursued along with efforts to create a new system of comprehensive global security. The components of such a system must be mutually supportive. Participation in it must be universal. The structure of such a system should be firmly based on non-violence.

> The new structure of international relations to sustain a world beyond nuclear weapons will have to be based on the principles of coexistence, the non-use of force, non-intervention in the internal affairs of other countries, and the right of every state to pursue its own path of development.

Throughout his speech, Rajiv Gandhi was speaking about non-violence and the role of use of force being limited to defence only. This is absolutely logical and essential for any progress toward a world free of nuclear weapons.

Nuclear weapons did not fall into the hands of human beings by mistake. It is true that at the time when nuclear weapons were invented, the designer, the military, the politicians and the public were not aware of all their dangers. But that does not mean that they got into their hands something they had not wanted – may be it was more than they had intended, but it was exactly what they wanted. And this became absolutely clear when the US and the USSR engaged in a competition of modernisation, even after all the dangers of nuclear weapons became known. So nuclear weapons were not invented accidentally. People were striving to invent more and more powerful, non-selective and massive weapons. This was the demand for the massive industrial wars. And this was the same logic which led to the concept of counter-value strikes, for which nuclear weapons are the ultimate ones. So nuclear weapons did not fall into the hands of human beings by mistake or unintentionally. Consequently, we cannot approach the ambitious mission to build a world free of nuclear weapons as a task to fix such a perceived mistake.

To build a world free of nuclear weapons, we must go into the very fundamentals of reliance on nuclear weapons. The starting points for this are:

- Rule of international law in international affairs and limitation of legitimate use of force for defence only.
- Improvements in regional security systems, in particular in the Middle East and South Asia.
- Correction of the conventional imbalances between major global and regional powers.

The comprehensive approach, which Rajiv Gandhi was speaking about and which is to ensure a fair and non-violent global security system, is not only more intellectually correct and consistent in comparison to the more narrow approach of doing something only about nuclear weapons, it is also much better tactically. A far more promising and realistic approach toward a world free of nuclear weapons is not to antagonise nuclear states that rely on nuclear as weapons of necessity, but to engage them in efforts to build a better and more secure world. ■

Approach to Nuclear Disarmament: Devalue to Discard

☐ **MANPREET SETHI**

INTRODUCTION

"Nuclear weapons capability is not necessary to peace, to security, to prosperity, to national greatness or personal fulfillment," said the president of USA at the time of the Indian nuclear tests in May 1998. And yet, there are few takers for the universal elimination of nuclear weapons.[1] The debate over nuclear abolition has always floundered on two basic issues – the *desirability* of achieving such a state; and the *feasibility* of doing so. Not many of those who matter are convinced that a nuclear weapon free world (NWFW) would be a better place to live in. They fear that such a world would, in fact, be more prone to wars in the absence of nuclear weapons that, in their perception, have been the keepers of peace. At the same time, there are several, including among those who question the desirability of universal elimination of nuclear weapons, who are sceptical that such a state can actually be achieved. Their presumption is based on two factors: one that the knowledge of how to make a nuclear weapon can never be wiped off from human memory. So, the risk of its being reconstituted by a delinquent nation cheating on its commitment to not acquire or possess nuclear weapons would always remain; second, because achieving an NWFW is a technically challenging proposition given the difficulties involved in dismantling the weapons, disposing off fissile material, arranging for its safe storage at secure sites, and crafting adequate safeguards to oversee the processes. The tenacity of the desirability and feasibility arguments is proven by the fact that the world has not been able to shake off the stranglehold of nuclear weapons for the last six decades despite opportunities to do so.

In most recent memory, one such opening came about in the early 1990s with the end of the Cold War. There were many who then opined that the demise of East-West confrontation would eventually deprive nuclear weapons of any utility and hasten their removal from national arsenals. Some even went to the extent of writing obituaries of war itself.[2] Optimism resulting from the fast developing cooperative relations between the US and Russia led one analyst to capture the spirit of the changing times in these words, "These are the days of hope, not despair... It is time to strip away the complex and arcane strategic theory of the Cold War and start from scratch."[3]

These prognostications, however, had underestimated how deep the roots of nuclear deterrence had penetrated over the years and the period of hope proved short-lived. So today, nearly two decades after the end of the Cold War, all nuclear weapon states (NWS)[4] continue to rely on nuclear weapons as a cornerstone of their national security for the "indefinite future." They have redrafted their nuclear doctrines to assign newer roles and missions for their nuclear arsenals. If earlier, the Cold War justified their existence, it is now the end of the Cold War and the emergence of two new threats – those arising from weapons of mass destruction (WMD) proliferation to "rogue states", in American parlance, and from the avowed interest of terrorist organisations in acquiring WMD – that vindicates their continued existence. In fact, an examination of recent trends in nuclear weapons thinking and policies of NWS indicates that each is maintaining its nuclear weapons capability to hedge against future uncertainties.[5]

The problem with this argument is that it raises risks of perpetuating an ever-widening circle of nations with nuclear weapons, materials and technologies. The world of nuclear weapons exists as a globalised entity, in cause and effect, linked by enduring ties of terror and response. Any development in the nuclear arena of one country has direct or indirect implications (through a cascading chain reaction) on one or more countries, eliciting a proportionate or disproportionate response, leading to more of the same. While it is conceivable that the international community may be able to devise a *modus vivendi* for a situation where many states have nuclear weapons, this dramatically increases the possibility of proliferation to non-state actors, a situation that is far more dangerous. Nation-states, as rational players and stakeholders in the international system are mostly expected to employ nuclear weapons as a 'deterrent'. Notwithstanding a high rhetoric over low nuclear thresholds by some NWS, no state can contemplate nuclear use without factoring in its own large-scale destruction. And this is expected to cast a constraining influence on the actual use of nuclear weapons for war-fighting. In contrast to this, for the non-state actor, the attraction of these WMD lies in their use. There would be little to gain and much to lose for a terrorist group to acquire the capability to mount nuclear terrorism, announce it, but not use it. Therefore, if a terrorist organisation was to acquire a nuclear weapon or even radioactive material, it should be expected to use it, not maintain it for existential deterrence.

The international community, led as always by the US in the nuclear arena, has employed a multi-pronged approach to grapple with the new challenges.[6] The governments, however, are still chary of considering universal elimination of nuclear weapons as a viable and sustainable way of handling the threat. The four retired US officials, through their two articles in the *Wall Street Journal*, have brought the focus on disarmament over the last couple of years, but serving decision-makers are yet to be seized of the idea. However, the one perceptible change that has come about with the emergence of new nuclear risks is that fewer people today doubt the desirability of an NWFW. Instead, the scepticism is now confined to the enormity of the challenge of making it feasible.

AIM OF THE PAPER

This paper presupposes the desirability of a state where nuclear weapons have been universally eliminated from national nuclear arsenals. It, therefore, concentrates on exploring the possible routes for reaching that state. This issue has been addressed several times earlier too and there is a distinct sense of *déjà vu* as one begins to outline an approach to disarmament. Strategic analysts, retired officials, and political leaders have in the past elaborated different approaches to nuclear disarmament — creating and sustaining a lower salience nuclear world, operational arms control, phased reductions, virtual nuclear arsenals, international control of nuclear weapons, phased elimination, nuclear weapon free zone creep, and time-bound framework.[7] The paper draws upon these ideas to configure its own approach to an NWFW. It identifies the major milestones where the world could take a breather after trudging along on a journey that will be long and difficult. Long because it would be practically impossible to bring about an NWFW overnight. The very act of dismantlement and physical destruction of the large arsenals accumulated in the NWS would take time. And difficult because the complexity of the task is exacerbated by the fact that it is intertwined with issues of state sovereignty, power play, national prestige, etc. A few caveats are, therefore, in order:

(a) An NWFW cannot come about instantaneously. The complexity of the task makes it impossible to expect that all countries would at some dawn see sense in arriving at a comprehensive treaty on nuclear abolition and do so in one go. It will have to be approached slowly, from different directions and with patience.

(b) An NWFW can be achieved if the agreed steps, however small, are taken with sincerity, backed by belief, sustained by passion, and supported by innovation. Any half-hearted measures that reveal lack of commitment would be harmful since they would reinforce the belief that it is an unfeasible proposition.

(c) An NWFW need not wait for all security-insecurity co-relations between countries to be addressed through a system of world governance in which all live in complete harmony before it can be realised. While a world that has renounced armed conflict as a means of settling disputes would indeed offer the ultimate security guarantee and remove justification for such weapons, the challenge lies in conceiving nuclear abolition in this real world inhabited as it is by sovereign states with conflicting national interests.

(d) An NWFW must, therefore, follow a mop-where-you-can approach to begin with. Small, doable steps taken along a range of issues to address concerns and build confidence will enable more difficult steps to be taken in the years to come. Therefore, the approach must be calibrated to suit present conditions and taken in as cooperative a manner as possible.

APPROACH TO DISARMAMENT

The paper strongly argues in favour of approaching the goal of an NWFW by progressively devaluing nuclear weapons and eventually delegitimising them. It is human nature to cling on to anything that is perceived to be valuable. In case nuclear weapons can somehow be made worthless, nations would be more open to giving them up. Therefore, a devaluation strategy that deprives the weapons of utility and renders them unusable through a series of measures can prepare the ground for their eventual elimination.

Devaluation steps may be taken along three trajectories: attitudinal – relating to mindsets; actual – relating to doable acts; and accessorial – relating to additional measures. These need not be sequential. While each would create an environment making it easy to take another, any one of them could follow the other, or even be pursued simultaneously. The purpose would be to soften the mud around the deep-rooted tree of nuclear deterrence so that when the states are ready, it is possible to uproot it with ease.

ATTITUDINAL DEVALUATION

Targeting Nuclear Belief Systems

Beliefs that deify nuclear deterrence for its war prevention capabilities ascribe greater and greater value to nuclear weapons. Therefore, it becomes necessary to attack belief systems that grant a larger than life image and role to nuclear weapons. Changing mindsets is imperative for devaluing nuclear weapons. Since the majority of the strategic decision-makers have believed in the undesirability and impracticality of nuclear disarmament, this belief has gone on becoming entrenched. In order to break the mould, the belief systems need to be strategically "attacked" and replaced by those more favourable to nuclear abolition.

History provides ample examples of institutions whose abolition was once thought to be unthinkable but which broke down under the pressure of changed belief systems. One example exists in the case of abolition of slavery. In fact, when the first demands for abolition of slavery were made, there were several who argued against it on the grounds that the economic and social systems would not be able to survive the drastic change and would collapse!! At the time, on December 1, 1862, President Abraham Lincoln, in his famous message to the US Congress, had said, "The dogmas of the quiet past are inadequate to the stormy present... As our case is new, so we must think anew and act anew. We must disenthrall ourselves."[8] This advice is most relevant today in the case of nuclear weapons. We have to disenthrall ourselves from their hold and if anything can do that it is the force of new ideas. As Victor Hugo had once said, "Nothing else in the world... not all the armies... are so powerful as an idea whose time has come." There is an urgent need for brave, new thinkers who can visualise a new world and inject the passion for it into the decision-makers. Administrations are known to resist change but if a sufficiently enlightened political leadership can coalesce to form a critical mass, then the tipping point for nuclear disarmament, at least conceptually, can come about very quickly. Every revolution in history has taken its leaders by surprise because as long as the old order exists, it tends to hide its fragility through elaborate measures. Its control then seems total and impregnable. But ideas can still sneak in and slowly snowball into a force capable of bringing about a profound change.

Redrafting Nuclear Doctrines to Restrict the Role of Nuclear Weapons

Altered belief systems would automatically prompt a reexamination and redrafting of nuclear doctrines. This would be an extremely important step since the role of a nuclear weapon is defined by a country's nuclear doctrine. Therefore, to first restrict the role to a narrow core mission, and then take even that away, a rewriting of national nuclear doctrines is required. Nuclear weapons were originally developed to deter the use or threat of use of other nuclear weapons. Gradually, however, countries detected in them the power to offset even conventional inferiority. It is this belief that leads to the weapon being considered as one "of the weak against the strong, as the only weapon that can counter the conventional superiority of the West."[9] Not surprisingly, therefore, smaller states with heightened threat perceptions, see value in nuclear weapons. Secondly, these weapons are also perceived as bestowers of prestige and status, as also the right to independent decision-making. In the words of the French Prime Minister Alain Juppe, "By acquiring nuclear capability, France was able to play, on the world scene, a role well above that justified by its mere quantitative significance."[10] These perceptions enhance the utility of nuclear weapons and motivate others to strive for them.

To a large extent, it was to remove the attraction of nuclear weapons as a leveller of conventional inferiority that the NWS had agreed to provide negative security assurances (NSA) to the non-nuclear states party to the NPT. Under the UN Security Council Resolution 984 of 1995, the NWS reiterated their commitment not to use or threaten to use nuclear weapons against non-nuclear countries. However, the contemporary US trend in favour of expansion of the role of nuclear weapons to cover threats from other WMD (chemical and biological weapons) and a move towards nuclear preemption undercuts the credibility of the NSA. It propels other countries to reassess their own need for such a weapons capability to address their threat perceptions. This belief could only have been heightened after the conduct of nuclear tests by North Korea and the manner in which the North Korea-USA nuclear equation has evolved since then. While the US has been able to get the Democratic People's Republic of Korea (DPRK) to halt, and declare, its nuclear weapons efforts in exchange for some incentives, it is also true that the presence of a nuclear weapons capability with Pyongyang, however small, has restrained American options. If more countries have to be stopped from perceiving a similar multi-role capability in nuclear weapons, then, pending their abolition, it is imperative to devalue the weapon by narrowing down its role.

Unfortunately though, today, the trend is just the opposite. All NWS propound that nuclear deterrence is still required as a principal pillar of their national security. Articulations such as these provide a positive feedback to nuclear deterrence and reinforce its hold over strategic thinking, making it a self-perpetuating phenomenon. It is this feedback loop that needs to be broken. For this, a reassessment of the nuclear doctrines to make nuclear weapons less attractive by restricting, and eventually obviating their role in international politics becomes imperative. Once this happens, the motivation to acquire them would substantially recede.

Investing Confidence in Multilateral Measures

The current mood on multilateral instruments is not very inspiring. The NPT is under stress. The Comprehensive Test Ban Treaty (CTBT) has been in a state of limbo from 1999 onwards. In 2002, once the US walked out of the Anti-Ballistic Missile (ABM) Treaty, Russia announced its abandonment of the Strategic Arms Reduction Treaty (START) II. Recent Russian pronouncements on the Intermediate Nuclear Forces (INF) Treaty have put it under a cloud even as the Fissile Material Cut-Off Treaty (FMCT) is still awaiting direction in the Conference on Disarmament (CD). All of these indicate a general trend away from arms control, particularly of the multilateral variety.

Instead, the US, normally the leader in these matters, indicates greater inclination for unilateral initiatives in an attempt to reduce redundant weapons while keeping the option of reversibility. This was best evident in the negotiations and conclusion of the Strategic Offensive Reductions Treaty (SORT) in 2002. The brief (only 500 words long) US-Russia bilateral treaty codifies cutbacks in a very loose and flexible fashion. It legally binds the two countries to a reduction of their nuclear warheads to a range between 1,700-2,200 by 2012 but allows each side to determine the manner, sequence and time-table that it would follow and also how the dismantled warheads would be disposed of. Also, the treaty contains no provisions for limiting delivery vehicles. Rather, it allows the two countries to retain as many delivery vehicles as they like – without exceeding the limits of START I, which itself will remain in force until 2009 only, and expire before SORT reaches its targets. SORT also does not include the elimination of MIRVed missiles, a step that START II had stipulated. Neither have the USA and Russia laid down any verification procedures to establish the veracity of the cutbacks.[11] In fact, a move away from verification appears to be the current American tendency. The draft on the FMCT proposed by the USA in May 2006 at the CD also steers clear of enforcing any verification mechanism, leaving it instead to National Technical Means to detect non-compliance.

An expression of lack of faith in multilateral initiatives generates international instability, lowers confidence levels and increases the value of nuclear weapons. China, for example, has used every opportunity to decry the US attempt at seeking "absolute security" by freeing itself of international obligations, even as it undertakes rapid military modernisation that would include an increase in its own nuclear arsenal, development of counter-measures and space-based capabilities, and thereby raises the concerns of others. In this process, then, international security and stability are the losers.

ACTUAL DEVALUATION

Accepting No First Use (NFU)

While limiting the role of nuclear weapons to dealing with only nuclear contingencies provides NSA to non-nuclear weapon states (NNWS) and devalues the weapon to some extent, accepting NFU takes the task further by removing the possibility of their use against NWS too. Therefore, adoption of NFU is a crucial step towards the eventual delegitimisation of nuclear weapons since it would involve an assurance from every country that it would not be the first to introduce nuclear weapons into a conflict. Since there will not be a first user, it would effectively mean no use of the nuclear weapon at all.

Of course, there are critics of the NFU who dismiss it as nothing more than a declaratory policy that would mean little once hostilities actually break out between two nuclear nations. Such criticism, however, tends to overlook two facts. Firstly, that the adoption of NFU automatically translates into a different nuclear force posture and deployment pattern that ensure that the promise of NFU is kept. Doctrines that ascribe a war-fighting role to nuclear weapons envisage 'first use' to retain the military advantage and adopt launch on warning or launch under attack postures. To undertake preemption, both sides need a large infrastructure in the form of command, control, communication and intelligence (C3I), early warning, defences, etc. NFU, on the other hand, frees a nation of all these requirements by basing its nuclear strategy on a retaliation policy only. It allows for greater response time for self and a more relaxed posture for the adversary. Therefore, acceptance of NFU would be accompanied by de-alerting, de-mating and de-targeting.

Secondly, the efficacy of NFU flows from the fact that declarations on the conduct of war do have relevance and cannot be summarily dismissed. These laws set certain limits to the levels of permissible violence and establish useful norms. For instance, the Geneva Protocol of 1925 outlawed the use of chemical and biological weapons and except for rare exceptions, the weapons have remained out of the realm of war-fighting. In fact, eventually, even if several decades down the line, it did become possible to capitalise upon the norm to formalise conventions outlawing these WMD. Following this example, an NFU commitment from all NWS could be expected to translate into a Nuclear Weapons Convention some time in the future.[12]

A universal NFU would be even more relevant as nuclear weapons reduce. With small nuclear forces, the temptation to launch a disarming first strike would be high because of the 'use them or lose them' compulsions. But an NFU posture would remove this temptation for self and the adversary. If the adversary is under constant fear that a nuclear strike is imminent, its own temptation to use nuclear force would be higher. The risks to stability and the possibility of nuclear war with different use doctrines are evident from Table 1.

Table 1

Nuclear Posture Country A	Nuclear Posture Country B	Nuclear Threshold	Possibility of Nuclear War
First Use	First Use	Low	Very High
No First Use	First Use	Relatively High	Low
No First Use	No First Use	High	Nil or Very Low

If all NWS were to accept NFU, then none would ever initiate a nuclear strike. Lengthening the fuse on their possible use would be a step toward downgrading the importance of nuclear weapons since military planners would have to discount their role in contingency planning. Over a period of time then, the weapons could fall into a pattern of disuse and lose their utility.

Banning the Use or Threat of Use of Nuclear Weapons

A step even beyond NFU, a treaty prohibiting the very use or threat of use of nuclear weapons would provide a meaningful way to delegitimise the weapon. A taboo against nuclear use already exists. If this can be legalised, it would further devalue the weapon. The force of such a treaty that binds nations to a pledge that nuclear weapons shall not be used and that any violation would invoke commensurate retribution and a total boycott by all the countries of the world would make these WMD absolutely impotent and useless. None would want to acquire a weapon that could not be used, not in war, and, hence, not as a deterrent either. Neither would the unique status that nuclear weapons are deemed to provide seem worth aspiring for. Even "rogue states" would no longer have any use for these weapons for fear of concerted and certain reprisals. Therefore, a total ban on the use of nuclear weapons would directly strike at the very roots of their utility.

In this context, it might be mentioned that a norm of nuclear non-use already exists. Widespread popular revulsion against nuclear weapons and widely held inhibitions on their use influence nuclear decision-making. This was evident even in the 1950s when an officer in the State Department's Bureau of Far East Affairs warned during the Korean War that "even though the military results achieved by atomic bombardment may be identical to those attained by conventional weapons, the effect on world opinion will be vastly different. The A-bomb has the status of a peculiar monster conceived by American cunning and its use by us in whatever situation would be exploited to our serious detriment."[13] Since then, the norm has existed but it has never been legalised.

In fact, the legality of nuclear use remains in dispute because the 1996 International Court of Justice Advisory Opinion had announced that while the use or threat of use is "generally" unlawful, it could not "definitively" conclude whether it would be unlawful in extreme circumstances of self-defence if the survival of the state was at stake. The ambiguity is, therefore, maintained. Meanwhile, the UN General Assembly has periodically considered resolutions stating that the use of nuclear weapons was contrary to the "spirit, letter and aims of the UN." Way back in 1961, it had adopted a declaration by a vote of 55 to 20 with 26 abstentions. The US and North Atlantic Treaty Organisation (NATO) had then opposed it, contending that in the event of aggression, the attacked nation should be free to take whatever action with whatever weapons not specifically banned by international law. India, for several years now, has proposed a "Convention on the Prohibition of Use of Nuclear Weapons." It seeks a multilateral, universal and binding agreement prohibiting the use or threat of use of nuclear weapons through an international convention. In the year 2000, however, the move was able to muster only 109 votes in favour with 43 against and 16 abstentions. Predictably, the P-5 opposed the resolution. The US hailed instead a step-by-step process that embraced unilateral, bilateral and multilateral measures.[14] Interestingly, Japan, for all its abhorrence of nuclear weapons, also found it prudent to abstain for reasons similar to those voiced by the US. Such country positions, nevertheless, underestimate the significance of actually arriving at a convention banning use of nuclear weapons to devalue them substantially.

In fact, this is even more urgent today because the norm of non-use or the taboo against the use of nuclear weapons could erode in the future for, at least, three reasons. One, because the memory of the horrors of Hiroshima and Nagasaki is fading. The new generation that has no first hand experience of the damage and destruction that a nuclear weapon can cause may not be able to visualise the implications of nuclear use. Secondly, with the induction of mini or micro nuclear weapons, the temptation to use them on the assumption that they would cause minimal collateral damage cannot be discounted. The convergence of the destructive power of small nuclear weapons and advanced conventional weapons, as in the case of fuel air explosives, would also erode the distinction between the two, heightening the possibility of a casual resort to nuclear weapons if they come to be perceived as no different from other weapons. Thirdly, there could be forces that have a vested interest in destroying the taboo. For instance, it is documented that even in 1953, in a discussion with the US president, John Foster Dulles had admitted that "since in the present state of world opinion we could not use an A-bomb, we should make every effort now to dissipate this feeling, especially since we are spending such vast sums on the production of weapons we cannot use." There is nothing to suggest that similar opinions are being or will not be, voiced in the future. In view of these, a ban on nuclear use would immediately serve to remove the value of nuclear weapons. It is no longer possible to make the nuclear genie disappear, but it is still possible to ensure that the genie of nuclear use is not allowed to come out of the bottle.

Halting Further R & D for Nuclear Modernisation

Besides reducing the existing nuclear arsenals, it is also imperative that a complete halt be put to further modernisation of nuclear weapons. This is particularly important given the emergence of a new trend toward improving the weapons in such a way as to make them "more user-friendly". In the US, certain laboratories have pushed for the development of mini-nukes of smaller yields. It is contended that small, low-yield nuclear weapons could be effectively used to destroy even a deeply buried or hardened underground facility without causing too much collateral damage.[15] In 2005, the Reliable Replacement Warhead (RRW) programme was initiated to explore making existing nuclear warheads last longer without diminishing their explosive power. This programme is euphemistically described as aiming "to improve the reliability, longevity and certifiability of existing weapons and their components"[16], besides also enhancing their safety and security and refurbishing the US nuclear infrastructure.

This refurbishment is the third leg of the new triad articulated by the US Nuclear Posture Review (NPR) 2001, and is expected to repair and replace nuclear weapons production and development facilities. Known as the Complex 2030 programme, for the year it is to be realised, it would enable the US to increase reliable warhead production capabilities while reducing actual numbers. In the process, it would produce not "a smaller Cold War era nuclear stockpile", but "capabilities appropriate for 21st century threats,"[17] because as was stated by Linton Brooks, administrator of the National Nuclear Security Administration (NNSA), in his testimony before the Senate Armed Services Sub-Committee, "Today's Cold War legacy stockpile is the wrong stockpile from a number of perspectives."[18] The old stockpile contains too many high yield weapons that offer little or no capability against hard and deeply buried targets, provide no way to limit collateral damage, and cannot destroy chemical or biological munitions. The RRW seeks to rectify these limitations by matching nuclear weapons to the new threats.

Whatever be the American justification for this exercise, a decision validating the importance of smaller, low-yield and more usable nuclear weapons sets off a dangerous chain reaction, with global ramifications. Russia, is already believed to be thinking along similar lines. In mid-December 2006, Putin called for the "development of cutting edge strategic weapons" and emphasised "quality over quantity".[19] Such developments ascribe value to nuclear weapons instead of taking it away from them. Until now, arms control has focussed on quantitative growth of arsenals. Disarmament must devise arrangements for controlling the continuous qualitative upgradation of nuclear and conventional weapons.

Restricting Delivery Systems

Long range delivery systems like ballistic missiles and strategic bombers expand the role and scope of nuclear weapons. It, therefore, would make eminent sense to put in place some kind of a universal restriction on delivery systems in a bid to devalue nuclear weapons. If the weapon is not effectively deliverable, then, it would be of little use and, hence, not worth aspiring for or stockpiling.

In this context, it could be suggested that the Intermediate Nuclear Forces (INF) Treaty, presently in force only for the US and Russia, should be expanded to prohibit deployment anywhere of land-based missiles between the ranges 500 and 5,500 km. A multilateralisation of the INF Treaty, meanwhile, offers a way of addressing the threat of errant missiles from small states, since most of them would fall within the range that shall be prohibited by a universalised INF Treaty.

ACCESSORIAL DEVALUATION

Crafting a Safeguards Regime

In order for the measures outlined above to be meaningful and sustainable, it becomes necessary to craft an effective and acceptable safeguards regime that can provide adequate guarantee that nations are complying with their commitments. Such a regime would be possible only through the adept employment of a combination of technological, diplomatic and statecraft skills. It would have to trudge a fine line because in order to be effective, it would have to be relatively intrusive and in order to be acceptable, it would have to be relatively limited. The organisational and institutional mechanisms that would perform these functions would also need to be thought through. Of course, the IAEA that has several decades of experience in safeguards is an option. But it would need to be infused with a fresh flow of funds and political powers to take on the additional burden of new safeguards, which would arise after a careful definition of items, activities, and facilities that would need to be monitored.

Joint R & D of Verification Technologies

States can be expected to renounce the nuclear weapon only when enough technical expertise is available to ensure adequate verification of the commitments undertaken to abandon nuclear arsenals. History illustrates that often changes become acceptable only when it becomes technically possible to prove something. For centuries, man continued to believe that the sun revolves around the earth. Those who disputed this belief were ridiculed or persecuted. Copernicus in his lifetime could not find acceptance for his theory. The belief changed only after the invention of the telescope and other astronomical instruments that could empirically prove that the sun was actually the centre of the universe. Applying the same logic to nuclear disarmament, it may be said that advancements in the field of surveillance and monitoring technologies, coupled with the research and development on how to deal with the nuclear materials obtained from dismantled weapons would help overcome the technical challenges to the feasibility of disarmament. If pursued as a multinational effort, it could help build deeper technical relations and lead to greater confidence in the process.

Dealing with Delinquents

In order to devalue nuclear weapons, it becomes necessary to put in place effective enforcement strategies that clearly lay down penalties for cheating or breakout. The guarantee of a collective and concerted international response could act as an effective deterrent to the clandestine development of nuclear weapons once all countries have committed to their non-use or abolition. A resolute response with sure penalties for any breach of trust would stem the temptation to go down the nuclear path or to use it as a lucrative bargaining chip. Only a collective endeavour to respond surely and united could effectively serve to devalue nuclear weapons. Of course, a determination of the exact nature of penalties that may be acceptable to all requires further investigation. But it is not a challenge that cannot be surmounted if the requisite political will exists.

CONCLUSION

The paper has identified a number of steps that could help to devalue the importance of nuclear weapons. These are largely based on common sense and logic. However, country positions are not always dictated by these two parameters. Ample evidence of this is provided by national positions, as revealed in the voting patterns on the different resolutions that continue to be placed before the United Nations General Assembly (UNGA) on the issue of disarmament.[20] The NWS and their allies tend to support resolutions listing the progress already made in the direction in terms of universalising the NPT, the conclusion of the CTBT, negotiations on the FMCT, reductions negotiated under the START process, etc. However, they oppose all resolutions seeking an actual delineation of steps to eliminate nuclear weapons in a time-bound or a phased manner.

Nuclear weapons can be abolished only when the belief systems behind their use and utility change. Existing norms of non-use can be institutionalised into legal regimes. Initially, it shall all have to begin in the mind and thereafter the possibilities would grow. The vision of an NWFW can be realised if it can draw upon new thinking at different levels: exemplary political vision and statesmanship willing to discard entrenched beliefs and accept new norms of inter-state behaviour; resolute military courage to shake off the crutch of nuclear deterrence and accept seemingly lower levels of security, especially for states that are conventionally weaker and see nuclear weapons as levellers of sorts; large doses of logistic innovation to deal with the discarded weapons; and an international system based on greater transparency, confidence and more participative institutionalised mechanisms.

Devaluing nuclear weapons would be an effective way to ultimately discard them. Complete elimination has often been dismissed as unattainable unless war as a means of settling disputes between states is first abolished and a state of general and complete disarmament is enforced through a system of world governance. Since the attainment of these prerequisites appear Utopian, nuclear disarmament too tends to be dismissed as unattainable. However, simple and easily implementable steps to devalue nuclear weapons prove that there are possible ways of approaching nuclear abolition. In this context, it is heartening to draw upon the experiences of outlawing other classes of WMD and activities without waiting for a comprehensive solution. In fact, it must be realised that nuclear weapons are in a class by themselves considering the indiscriminate, comprehensive, and long-lasting destructiveness that they can cause to plausibly destroy all civilisation. For the sake of the future of mankind, then, it is imperative that these steps be taken. ■

NOTES

1. The terms nuclear abolition, elimination and disarmament are used interchangeably in the paper to mean a world that has been rid of all nuclear weapons.

2. Barry Blechman and Cathleen Fisher of the Stimson Centre argued in the early 1990s that technological diffusion and economic interdependence had created a world order in which states shared so many common interests as to delegitimise the "very idea of using military force in the settlement of disputes." As quoted in Robert A Manning, "The Nuclear Age: The Next Chapter," *Foreign Policy*, Winter 1997-98, p. 76

3. Lincoln Wolfenstein, "End Nuclear Addiction", *Bulletin of Atomic Scientists*, vol. 47, no.4, May 1991.

4. The term nuclear weapon state is normally used to designate the five nuclear powers accepted under the NPT as those that had detonated their nuclear devices before January 1, 1967. However, in this paper, the term is used in a more loose sense to denote any state with a nuclear weapons capability.

5. For a detailed examination on this, see Manpreet Sethi, "Current Trends in Nuclear Weapons Thinking and Strategies," in Jasjit Singh, ed., *Asian Defence Review*, 2007 (New Delhi: Knowledge World, 2007), pp. 65-98.

6. Strengthened export controls as mandated by UNSCR 1540, the Container Security Initiative, Proliferation Security Initiative, Global Nuclear Energy Partnership,

Global Strike Plan are some of the US-led measures to address the proliferation challenges of the 21st century.

7. For an excellent exposition on the nuances of the various approaches, see Darryl Howlett, Ben Cole, Emily Bailey and John Simpson, "Surveying the Nuclear Future: Which Way from Here?" *Contemporary Security Policy*, vol.20, no.1, April 1999, pp. 5-41.

8. Jerome B. Wiesner, "A Militarised Society," in Len Ackland and Steve McGuire, eds., *Assessing the Nuclear Age* (Chicago : Educational Foundation for Nuclear Science, 1986), p. 227.

9. Robert G. Joseph, "Nuclear Deterrence and Regional Proliferators," *Washington Quarterly*, vol. 20, no. 3, Summer 1997, p. 168.

10. Speech delivered by French Prime Minister Alain Juppe at the Institut des Hautes Etudes de Defence Nationale, Paris on September 7, 1995, and as reproduced in *Strategic Digest*, April 1998.

11. The onerous and complicated task instead has been left to a Bilateral Implementation Commission that is expected to meet at least twice a year to deal with issues of compliance.

12. At present, only two countries – India and China – accept NFU. The latter, however, does not offer an unconditional NFU in contrast to the one spelt out by India that is clear and unambiguous. China maintains that its NFU commitment does not apply to its own territory or territories that it claims as its own. Hence, the ambiguity regarding the possibility of Chinese nuclear weapons in a conflict over Taiwan, or certain portions of Arunachal Pradesh, an Indian state to which China lays claim.

13. Nina Tannenwald, "The Nuclear Taboo: The United States and the Normative Basis of Nuclear Non-Use," *International Organization*, vol.53. no.3, Summer 1999, pp. 433-468

14. See "Appendix: Summary of Resolutions," *Disarmament Diplomacy*, <http://www.acronym.org>

15. Theresa Hitchens, "The US Nuclear Debate: Issues of Concern," *BASIC Paper* no. 35, February 2001.

16. Jonathan Medalia, "Nuclear Weapons: The Reliable Replacement Warhead Program," *Congressional Research Service*, May 24, 2005, p.6.

17. As cited in Walter Pincus, "Plan to Study Nuclear Warheads Stirs Concern," *Washington Post*, April 6, 2006.

18. Statement as cited in Stephen I. Schwartz, "Warheads Aren't Forever," *Bulletin of Atomic Scientists*, vol 61, no. 5, September/October 2005. pp. 58-64.

19. "Russia Against US MD Plans for Europe," http://www.spacewar.com, December 27, 2006.

20. In the year 2000 (as also every year on an average), nearly four dozen such resolutions appeared before the UN. However, an examination of the voting patterns revealed the rather predictable positions of each country.

India's Role in Nuclear Abolition and Cooperative Security: Call for a New Pragmatism

☐ **GARRY JACOBS**

FIFTY YEARS ago, the World Academy of Art and Science was established at the initiative of eminent scientists such as Einstein, Oppenheimer and Rotblat who were deeply concerned by the destructive power of the nuclear genii they had helped release from within the atom. The logical imperative that compelled the Allies to develop a weapon of unparalleled destructive capacity for self-defence was transformed by a perverse logic into the greatest single threat to the security of the Allies and the entire human race. Since then, the Academy has dedicated its efforts to addressing a wide range of issues regarding the application of knowledge in world affairs, but none of greater concern than the continued existence and proliferation of nuclear weapons.

Although the threat of global warming evokes greater public and political concern today, nuclear weapons present the most imminent danger and their abolition must be the highest priority. This view also arises from an imperative logic. Nuclear weapons are the lynchpin that block global progress on a whole range of political, social and economic issues crucial to the future of humanity. The imperative to fully and permanently eliminate the dangers of nuclear weapons is the same imperative that requires and compels responsible governments today to ensure food security for all human beings, to promote prosperity for all nations, to insist on democratic principles of governance both nationally and internationally, and to protect the earth against impending environmental catastrophe. These diverse issues are bound together by an invisible knot and none can be satisfactorily addressed in isolation from the others, yet there is a natural order and sequence to their accomplishment that is founded on, and begins with, the abolition of nuclear weapons.

RISING EXPECTATIONS AND GROWING DISILLUSIONMENT

The Allied victory in World War II generated rising expectations of a coming age of peace, democracy and prosperity, expectations that were partially fulfilled by the rapid economic development of the West but grossly betrayed by the commencement of the Cold War. Ironically, the war that was fought to replace the rule of power with the rule of democracy led to the founding of an undemocratic United Nations. The war that was fought by unprecedented militarisation to abolish the threat of future wars led to a precarious militarised peace and armaments race that posed greater threats to human security than all the wars of the past. The bright hopes of 1945 were blackened by the emergence of military blocs and the proliferation and stockpiling of an inconceivable and unconscionable arsenal of nuclear weapons. Mutually assured destruction (MAD) was not only an apt description of the defence doctrine of those times, it was also a fit description of the mentality which conceived and perpetuated it.

Saner minds on both sides understood the irrationality of this doctrine. Harlan Cleveland, a past president of the academy, who served as assistant secretary of state under President Kennedy, described to me many years ago how close we came to reaping the harvest of that madness during the Cuban missile crisis. He also related the efforts he took afterwards as US ambassador to the North Atlantic Treaty Organisation (NATO) under President Johnson to convince America's European allies that nuclear weapons were militarily unusable. More recently, Robert McNamara, US defence secretary during that same period, has passionately argued the same case within the academy and before the public. Yet, in spite of that clarity of mind, still the imperative logic of the past has prevailed for decades and the proliferation of weapons multiplied.

The nuclear Non-Proliferation Treaty (NPT) might have been the most important doctrine ever written, had the intention of its authors ever been to fulfill the obligations imposed on its signatories. The hostile political climate during the Cold War may have made fulfillment of these obligations impractical; but the fact is, as McNamara has disclosed, that it was never the intention of the nuclear powers to fulfill their obligations under Article VI of the treaty to abolish their stockpiles of nuclear weapons. It was rather their intention to prevent others from acquiring the inordinate means of destruction that the five nuclear powers already possessed, to set in stone and legitimatise a nuclear apartheid world. The doctrine of the NPT is containment. It is bound to fail in achieving that objective because in the absence of fool-proof mechanisms to prevent proliferation, it invariably rewards countries that manage to acquire nuclear weapons.

MISSED OPPORTUNITIES

From the very inception of the NPT, India has taken a clear and consistent position on the need for total abolition of nuclear weapons, a position forcefully proclaimed by Rajiv Gandhi during his address to the UN in 1988. The timing of his initiative was highly significant. He perceived what few had been able to foresee, even a few months before, that the opportunity was emerging for the most dramatic breakthrough in international relations since the end of the war. At the same time, he also feared that, in the absence of a breakthrough on nuclear disarmament, India would not be able to withstand the logical imperative and public pressure to declare and demonstrate its own nuclear weapons capabilities. Ten years after his initiative, that fear became a reality at Pokhran. The question before the participants in this conference and before the Government of India today is whether his vision can be made a reality and whether it can be done now.

Four days before the US Senate ratified the Nuclear Test Ban Treaty in 1963 and just two months before he was assassinated, President Kennedy addressed the United Nations to affirm his commitment to strengthen global institutions and work for a rapid end of the Cold War. Just one year after Rajiv Gandhi's speech to the United Nations, the Berlin Wall was brought down, the Cold War ended and he too was assassinated. It is useless to speculate as to what might have been achieved had their deaths been averted. But it is vitally important to recognise that both of them perceived an opportunity for a breakthrough, which others failed to perceive.

The end of the Cold War did become a reality, but through a failure of leadership on all sides, the world has not been able to capitalise on the opportunity offered by that occasion. In the absence of serious initiatives to affect a radical breakthrough, the momentum of past events perpetuated business as usual. The relief and joy which everyone felt at the end of confrontation was not converted into strategic or organisational change to rid the world of the ever-dangerous nuclear stockpile accumulated over five decades. It is a basic truth of social existence that the organisations we establish in order to serve us, develop a life of their own and often compel us to serve them instead. The military industrial complex developed to defend the free world during World War II is true to that principle. Failing to comprehend that the weapons themselves possess an inherent force for their own utilisation and that the industrial complex that thrives on the arms race will not idly stand by while lucrative commercial interests are abandoned, the world failed to capitalise on the courageous initiative of Gorbachev and Reagan, and the timely proposal of Rajiv Gandhi.

The International Commission on Peace and Food (ICPF), of which I was member secretary and Jasjit Singh was one of the most valued contributors, conducted its first meeting in Trieste, Italy, just weeks before the end of the Cold War. The establishment of the commission was an Indian initiative conceived by the Mother's Service Society, Pondicherry, in 1987, at a time when few people believed that the Cold War could be ended within two years. The next few years were times of a heady euphoria when the prospects for a truly global peace seemed well within reach. Yet after the ratification of Strategic Arms Reduction Treaty (START) 1 and Strategic Offensives Reduction Treaty (SORT), the momentum of progress quickly petered out. Although political capital is frequently made out of the fact that the number of warheads has declined dramatically from its unfathomable peak of 65,000 in 1985, political capital has not translated into greater global security. The reduction in warheads is of no practical significance in view of the approximately 10,000 weapons that are still active. Thousands of those warheads remain on alert status. Three more countries have joined the nuclear club. The US has upgraded the role of nuclear weapons in defence strategy. New nuclear weapons are under design. After declining dramatically for a decade, world military expenditure measured in constant dollars has been rising since 1998 and is back to the peak levels of the Cold War.

The dangers of nuclear proliferation are probably greater than at any time in the past. And to cap it all off, there is a serious intent to extend the arms race into outer space. Eighteen years later, we have still to deal with the consequences of that failure. All this has occurred in spite of the fact that war between nuclear powers has become unthinkable. Ironically, with few exceptions, the countries that most depend on nuclear weapons for their security are the ones that need them the least and are most threatened by the very existence of these weapons. The prelude to World War II offers stern lessons to the world regarding the dangers of acquiescence and appeasement. Those lessons counsel us not to make the same mistake today by acquiescing in the prolongation of an unstable and untenable nuclear status quo.

CALL FOR PRAGMATISM

Pragmatism is the prevailing wisdom governing international discussions on the issue of nuclear abolition. In the name of pragmatism, emphasis is placed on the possible rather than the desirable, the realistic rather than the idealistic. This has led to the formulation of a comprehensive list of possible next steps, including measures such as the ratification of the Comprehensive Test Ban Treaty (CTBT) and the Fissile Material Cut-off Treaty (FMCT). No doubt, these measures are natural first steps and vitally important to the ultimate abolition of nuclear weapons. Pragmatism focusses on the incremental and the immediate in preference to quantum leaps and the long term. It counsels us to moderate our expectations based on the implicit assumption that the scope for progress in the short term is relatively modest.

But the Cold War was not brought to an end gradually by a series of incremental steps. It was precipitated by rapid and radical action that took the entire world by surprise. The Cold War was not ended by pragmatism. Vision, courage and determination ended it.

There are even occasions in which an exclusive concentration on incremental change not only fails, but can be counter-productive. Even if the CTBT and FMCT were unanimously adopted today, there is no assurance that they would ever lead to nuclear disarmament. They could just as well serve to perpetuate the status quo by removing the most serious incentive for the nuclear powers to give up existing weapons – the threat that still other countries may acquire them. In such instances, the only truly pragmatic approach is to view an issue comprehensively and recognise the need for a quantum shift rather than mere incremental improvements. The issue of nuclear weapons is one such. Pragmatism compels us to examine the issue of nuclear abolition in the wider framework of global cooperative security.

COMPETITIVE VS COOPERATIVE SECURITY

The term cooperative security has been liberally applied and misapplied in so many different ways that usage of the term requires clarification. The Allies applied a cooperative security framework during World War II to wage war against the Axis powers. We need a cooperative security that is united for peace, not war. During the Cold War, the two blocs each adopted an exclusive version of cooperative security to prepare for war against the other. Today, we need a cooperative security system that is inclusive rather than exclusive.

True global cooperative security is a concept that has never been tried. The prevailing security paradigm of the past is a competitive approach to security, one in which each nation or bloc seeks to enhance its own security by building offensive and defensive military capabilities that decrease the perceived or real security of other countries that are not included in the system. As the International Commission on Peace and Food pointed out in their report to the United Nations, "The push of each nation or group for unlimited security through military power is inherently destabilising, since it inevitably increases the insecurity of other nations and compels them in turn to increase military preparedness."[1]

One of the greatest inherent fallacies in the current approach to both global security and nuclear non-proliferation is that it ignores the right and responsibility of each nation to provide for, and ensure, its own legitimate security needs in the absence of effective alterative mechanisms at the international level. This understandable and inevitable compulsion undermines the current approach of the original nuclear powers, which seek to maintain military dominance to ensure their own security without addressing the legitimate needs of other nations. This fragmented approach to global security is essentially flawed and doomed to failure. The only pragmatic approach to nuclear non-proliferation is to evolve a security framework that addresses the needs of all countries. In spite of international outrage over the acquisition of nuclear weapons by India, Pakistan and North Korea, the fact is that all three nations have raised their status internationally by so doing and provided an attractive example for other countries to emulate. That is the logical imperative of competitive security.

BASIS FOR COOPERATIVE SECURITY

Pragmatism may rebel against broadening the focus from nuclear non-proliferation to cooperative security, but I argue that anything less than a comprehensive approach is unrealistic and will only lead to further proliferation in future. The case of Iran is indicative.

There are times when what appears as the most difficult path proves to be the path of least resistance. The end of the Cold War was a sudden, astonishing and multi-dimensional fulfillment of global human aspirations. The end of military confrontation between the superpowers and subsequent dissolution of the Warsaw Pact was only one aspect of that remarkable accomplishment. Simultaneously, it brought about the political liberation and self-determination of previously suppressed nations. It unleashed social revolution within these countries and led to a rapid surge in living standards throughout the region. But it did not end there. The end of the Cold War also created conducive conditions for four still more remarkable achievements, whose immense importance for global human security has yet to be fully recognised.

First, it facilitated and accelerated efforts of the nations of Europe to build a supra-national union of sovereign states – on a continent that had been wracked by national rivalries and incessant warfare for the previous 500 years – leading to the establishment of the European Union (EU) in 1993 as a political and economic community. The concept of a united Europe long preceded these events, but its realisation awaited the right environment. The expansion of the EU to include 27 member countries was difficult to even imagine a mere 20 years ago.

Second, and closely related to the first, was the establishment of the European Monetary Union in fulfillment of an idea first proposed in 1929 and revived in the 1970s but not acted upon. It is highly significant that the debate acquired serious intensity only in the year immediately preceding the end of the Cold War and that the first concrete step of abolishing exchange controls occurred nine months after the fall of the Berlin Wall. The creation of the

European Central Bank occurred just ten years ago to this month. A year later, the euro was born, the first transnational currency of global significance.

Third, the end of the Cold War made possible the establishment of the World Trade Organisation, the integration of China within the global economy and an unprecedented growth in global prosperity. World trade in merchandise and services has multiplied four-fold since 1990 and has doubled in the past seven years.[2] According to McKinsey, total global financial assets have increased from $12 trillion in 1980 to around $180 trillion today. Global forex reserves have risen nearly ten-fold since 1990, from $790 billion to $7.5 trillion in early 2008. India, which was down to its last few billion dollars in foreign exchange reserves during the first Gulf War, now has forex reserves of $316 billion.

Fourth, the end of the Cold War paved the way for the birth and phenomenal growth of the first human-centered global social organisation – the world wide web. Although the connection has not been widely recognised, the two events are inextricably interrelated. Without an end to East-West confrontation, it is inconceivable that an open global information system such as we have today could have evolved. Security concerns – whether reasonable or unreasonable – would have militated against it. The web is not merely an unprecedented achievement in itself, but it has been the spearhead and foundation for a movement of global economic and social integration that is too complex and comprehensive even to accurately describe. Without the web, India would not be the global information technology (IT) powerhouse it is today. Without the web, global social and economic integration could never have attained its present scale. Without the web, the economies of India and China – representing 40 per cent of the world's people – would not be growing at upwards of 9 per cent and creating sufficient jobs for their growing populations.

My purpose in detailing these developments is to emphasise the fact that, not only was the end of the Cold War unforeseen, but its impact and consequences have been unimaginable. A similar potential lies concealed and barred by the continued existence of nuclear weapons. The eradication of all stockpiles will not only eliminate the real and present danger of proliferation and ever-present threat of their usage, a possibility that we discount at our peril. More importantly, it will open the floodgates for other changes that we are barred from effectively dealing with before the nuclear lynchpin has been removed. Highest on the list of those changes are the elimination of war, global prosperity, and effective action to reverse global warming.

GLOBAL COOPERATIVE SECURITY SYSTEM

In recent decades, it has become increasingly apparent that a narrow conception of security focussed on military action is far too limited and inadequate to address human needs. The current worldwide crisis concerning rising food prices and rising energy prices is illustrative of how important these other needs are for the security and maintenance of global society. Recognition of the potentially catastrophic impact of climate change on sea levels and weather patterns has awakened public concern around the world and mobilised nations to action. But none of these issues can be effectively addressed in isolation from the others. They are inextricably linked to one another. A perception of that linkage prompted Indira Gandhi to point out at the Stockholm Conference on Environment and Development in 1972 that poverty is the greatest source of pollution in the world. Thirty-five years later, the abolition of poverty has acquired reality and is in sight.

The issue of climate change is unique in being the one serious threat to human security which cannot be addressed effectively by unilateral, bilateral or even multilateral action. Unless all the major economies of the world join together, the recalcitrance of some will be borne by all nations. Concerted action on climate change demands global action and necessitates the development of fully representative and participative global institutions. The trend toward unity began long ago and has been fostered by development of more than 100 international organisations over the past six decades. But the present structure is flawed and inadequate to serve the world's needs. Global warming is nature's warning that the task of evolving effective institutions for global governance must not stop halfway.

As there is a linkage between poverty and the environment, there is a very strong linkage between peace and prosperity. The need to ensure full employment was a fundamental objective of the Allied nations after World War II. In 1944, President Franklin D. Roosevelt declared full employment a basic human right. Full employment was one of the founding objectives of the Organisation of Economic Cooperation and Development (OECD). In 1992, ICPF submitted a full employment strategy to the Government of India to create 100 million jobs within ten years, and the strategy was formally adopted.[3] A decade later, India became the first major nation in the world to pass legislation guaranteeing employment and to introduce a national programme to fulfill that objective. Absence of employment opportunities is a root cause of social violence and a breeding ground for terrorism. Guaranteed employment, coupled with elevated higher standards of education, is the surest social foundation for peace.

Addressing the wider concept of cooperative security requires a comprehensive approach to global issues. Enforcement of the NPT, and ratification of the CTBT and FMCT should be vigorously pursued, but there is no assurance that they would represent significant steps toward global security. Eradication of nuclear weapons, on the other hand, would definitely represent such a step that opens up far greater possibilities for world peace and global development.

Social progress is a complex multi-dimensional movement that proceeds simultaneously on many different fronts, each supporting and reinforcing the others and adding to the overall momentum. Thus, the soundest and most pragmatic approach to addressing the issue of nuclear proliferation and the abolition of nuclear weapons is to move forward simultaneously on multiple fronts. A truly viable global cooperative security system must include appropriate material, social, political and military mechanisms.

World Peace Army

The most essential requirement for establishing a global cooperative security system is to put in place a viable alternative to national or regional military forces, which are invariably perceived as threats by other nations. The framework proposed by ICPF for this purposes is a World Peace Army (WPA), membership in which is open to all democratic nations that eschew resort to all forms of military aggression against other nations. Membership in the WPA would require contribution of financial, technological and human resources toward a common military force similar in principle to the proposal put forth for a European Army in 1998 and renewed by Germany late last year. In return for these commitments, members would be provided unequivocal security guarantees by WPA and irrevocable assurances of support in the event of armed attack by any nation. To quote from ICPF's report:

A World Army could consist of an international peace force that would unconditionally guarantee the security of its members against external aggression based on the following provisions:

- Membership in the World Army would be voluntary and open to all countries, provided that they have, and maintain, democratic, multi-party political systems. Since membership is not exclusive, it could be merged at any time with other like-minded organisations such as NATO, or be integrated into a UN military force when the necessary changes in the UN political structure have been made.

- As a condition for membership, each country would, by legal enactment, forego war as an instrument of policy and undertake not to commit any act of external aggression against any other member or non-member for any reason whatsoever. Any violation would be grounds for immediate expulsion.

- Members would agree to contribute an assessed sum of money, equipment and military personnel towards the maintenance of a standing military force under the command of a centralised military leadership. In addition, the peace force could be granted the right to recruit personnel directly from member countries.

- Members would also agree to limit their own overall military spending within norms fixed by the organisation.
- Members would be banned from possession of nuclear weapons and ballistic missiles. Existing weapons would be destroyed or turned over to the peace force.
- As in NATO, members would agree by treaty to consider an attack on one member as an attack on all. The peace force would intervene automatically and unconditionally – defensively and, if necessary, offensively – to protect the sovereignty and international borders of any member country, provided that it conforms to the rules under the charter.
- The organisation could also assist members in fighting drug trafficking and terrorist activities.

The benefits of this cooperative peace-keeping mechanism would be manifold:

- Since the charter would bind members and the alliance to eschew the use of force and aggression for any reason, it would represent a stabilising, non-provocative, non-offensive defence system. Its charter could also include specific provisions for close association with other international forces.
- Its combined strength and technological capabilities would exceed those of any of its members and constitute a substantial deterrent to aggression against any member country.
- It would significantly reduce the costs of security for members, possibly by up to 50 or 75 per cent, without compromising their legitimate security needs. The more members it included, the lower the defence expenditure required by each, both because the collective force would be larger and because the number of potential adversaries would be reduced proportionately.[4]

World Central Bank and World Currency

The subprime mortgage crisis, which struck the US financial system last year and is still playing itself out, has already accounted for direct losses by financial institutions of more than $500 billion, half of it by non-US financial institutions. It has also resulted in a huge destruction of financial assets on the world's stock markets – in January 2008, the Bombay market plummeted by more than 25 per cent from its peak in December 2007. It has also affected the growth prospects of the major economies of the world. Although it is too early to accurately access the magnitude of the losses incurred, the total loss to global gross domestic product (GDP) and financial wealth amounts to trillions of dollars. In spite of widespread publicity and debate regarding the causes and remedy for this crisis, little attention has been focussed on the single most important factor: the absence of effective international institutions to monitor, manage and govern the global financial system.

Money is no longer a national institution governed by nation-states. Money has gone global. More than $5.3 trillion in traditional and over-the-counter foreign exchange transactions take place every day in pursuit of short-term profits, wherever local conditions appear most attractive. At the same time, the ownership of investible assets has changed markedly. In 1990, foreign investors owned less than 10 per cent of equities worldwide compared with 25 per cent now. In the USA, foreign investors now hold about 60 per cent of all US Treasury securities, compared to 20 per cent in 1990. They also own 25 per cent of all US corporate bonds and 12 per cent of corporate stocks.

We live in the Wild West of global finance, akin to the period in American history when land was free for the asking or taking, and lawlessness reigned on the frontier. Historian Paul Johnson has depicted the opening of the American frontier as a crucial turning point in world history and the initial condition for America's rise to world leadership, but the immediate consequences of the frontier for other nations were relatively muted. The same is not true of money today. The global monetary frontier affects the economic and financial well-being of every nation in the world. Its governance cannot be left to market forces or chance. Yet, today, there is a vacuum at the top of the international financial world akin to the situation in the USA during the Panic of 1907, which ultimately led to the establishment of the US Federal Reserve. The world cannot afford a financial panic of global proportions. The stakes for global security are far too great.

More importantly, the phenomenal success of the euro – its 50 per cent appreciation against the dollar in the past six years – indicates the enormous potential benefits that can be obtained by introducing a truly global monetary system based on a global currency. The euro was originally adopted as the single currency by 15 countries. Today, it is also used by nine other countries, bringing a total of 500 million people under the euro umbrella. The value of a currency is directly related to the size of population using it. Money represents productive power. The greater the productive power of the region in which money circulates, the greater the overall value of the currency. If the expansion of the euro from its original 15 members has led to a 50 per cent increase in its value, imagine what would be the total economic benefit to the world of adopting a common global currency utilised by six billion people.

Progress toward global economic integration has advanced by the establishment of global standards with regard to technology, communication, transportation, manufacturing quality, financial systems and reporting, accounting principles, copyright laws, etc. The establishment of a common currency is to introduce a uniform, stable monetary standard for the transaction and measurement of financial transactions globally. This initiative will have the same magnitude of impact in the field of economics as the introduction of a global technical standard for the internet – HTML – had for the birth and explosive growth of the web.

Democratisation of the United Nations

The current undemocratic structure of the UN system is in direct opposition to the ideals it was established to uphold. The concentration of power among the five permanent members of the Security Council, most especially the veto power, undermines any claim of the UN to being democratic and representative. However justified or inevitable that structure may have been in the immediate post-war period, there can be no justification for perpetuating it further. At the time the UN was founded, more than fifty nations were still under colonial rule. Declaration of their rights to sovereignty and self-determination was understandably the highest priority. But more than six decades after its founding, a crude definition of national sovereignty which ignores the nature of national governments can no longer be justified. True sovereignty resides in the rights of the people, of every individual, for freedom and equality. Therefore, ICPF proposed in its report that a democratic form of government should be made a minimum requirement for continued membership of states in the UN system. It is self-evident that such a requirement must be mandatory for any nation to claim a legitimate right to a seat on the Security Council. A time-bound schedule should be adopted to impose this as a minimum requirement for UN and Security Council membership, failing which membership rights should be suspended.

INDIA'S ROLE

All social change requires the generation of a force of requisite intensity. Therefore, pragmatists focus on how to leverage the available political will and bring it to bear for incremental progress. But the will of national political interests and poorly informed public opinion is not the only force bearing on this issue, nor is it the most powerful and compelling. There is also a force of social evolution, a force of irresistible intensity and inevitability that is directing and driving the world's development in a specific direction. That evolutionary force is not a mere theoretical possibility. It is making itself felt in countless concrete ways today. It is the force behind the rising expectations of people everywhere. It is the force behind the rapid emergence of the internet as the first truly global system. It is the force behind the globalisation and integration of business and finance. It is the force behind global concern about climate change. Pragmatism also compels us to take cognisance of this force, rightly comprehend its direction and mode of action, and align overseas with its objectives.

The final refuge of pragmatism is to blame the failure of progress on the absence of visionary leadership. So it is claimed that if only Churchill, Roosevelt or Nehru were alive today, things would be different. But the 21st century is no longer the age of individual heroes. It is the age of impersonal institutions. Perhaps no individual leader, no matter how great, has the power to move the world today; but a government of India's stature does possess that power when it speaks directly to the aspirations of humanity and lends itself as an instrument of the evolutionary force.

India is uniquely endowed with the moral authority and legitimacy to act effectively on the issue of nuclear abolition. It has maintained a consistent opposition to nuclear weapons for half a century. It is the only nation to have won its independence by non-violence. Throughout thousands of years of its history, it has never committed an act of aggression against another country. It is, perhaps, the only land where a reigning Emperor, King Ashoka, renounced the very principle of war. It is the world's largest democracy. It is destined to soon become the world's most populous nation and one of the reigning economic superpowers. It is also the only major nation that has displayed the vision to guarantee employment, which is the essential economic foundation for the elimination of war and violence. More than any other nation, its spiritual culture is founded on adherence to a principle of truth that transcends petty self-interests. India is fully endowed and qualified to take up this challenge, if it chooses to do so.

Proud Indians, eagerly awaiting the day when their nation will be invited to join the permanent members of the UN Security Council, may hesitate to act in a manner they perceive may forestall that great event. But I would argue, on the other hand, that leadership on this issue will establish beyond doubt India's legitimate right to that status. Even now, there is no inherent logic and legitimate justification for India's omission from the UN Security Council (UNSC), any more than there is justification for an undemocratic UN or unfair NPT. As a nation that contributed more than a million soldiers to the Allied cause, India was fully qualified to assume a seat on the Security Council as soon as it achieved independence in 1947. Its claim to ascension was certainly as valid as any claim by the Chinese that led to their entry in 1960. Probably everyone attending this conference concurs that India has a rightful claim to UNSC permanent membership. But not all may realise that the very same logic that justifies the status quo on nuclear weapons is also applied to justify the status quo of an undemocratic and unrepresentative UNSC. By the same irrepressible logic of the past, India, which is now a thriving democracy, a major military power and a recognised future economic giant, remains outside the global political and security system, striving to get a foot in the doors of international power. The world is waiting for leadership on this issue and India should heed the call of destiny. ∎

NOTES

1. See *Uncommon Opportunities: Agenda for Peace & Equitable Development* (Zed Books, 1994), p. 40, or http://www.icpd.org/UncommonOpp/CHAP03.htm.
2. Based on WTO data in current US dollars, total world trade was approximately $2.4 trillion in 1980, $4.3 trillion in 1990, $7.9 trillion in 2000 and $17.1 trillion in 2007.
3. n.1, pp. 122-124, http://icpd.org/UncommonOpp/CHAP05.htm
4. Ibid., p. 45.

Regional Restraint, the Chemical Weapons Disarmament Experience and a Nuclear Weapon Free World: Can Asia Lead?

☐ **RORY MEDCALF**

I WANT to cover several interconnected issues today, but particularly how Asian powers might work creatively to reduce nuclear dangers in their region as a building block to global disarmament. The history of chemical weapons (CW) disarmament and delegitimisation offers one window onto this.

My basic thesis is that nuclear arms control and disarmament diplomacy has not caught up with the rise of Asia. By "the rise of Asia", I refer in this context to several things:

- The increased wealth, power and influence of Asia, especially the region covering East Asia plus India.
- The expansion and the promise of Asian regional cooperation, through the creation of new multilateral settings in the region.
- The increased diplomatic heft that Asian countries can wield, both through these forums, and, were they to work together, on the world stage.

The course of action I propose may not be the only or the best way forward for Asia on nuclear arms control, but it should at least prompt thinking as to how the rise of Asia can better be utilised in pursuit of nuclear security.

It is fair to say that, until the current conference, the latest international push for nuclear weapons abolition has been most prominent in Western states: e.g. US (second track), UK, Norway. It is heartening, therefore, that India, in keeping with its traditions, is putting its growing weight behind this quest. And India's reemphasis of its nuclear disarmament aspirations is very much in the spirit of the vision of Rajiv Gandhi, who was determined to see Asia take a lead on this front.

But much of the recent work on these issues has primarily involved familiar Western faces. Yet, these days, my country Australia's strategic horizons are primarily Asian. And it is in Asia where the choices between nuclear restraint and risky nuclear postures, and the choices between accepting proliferation or pursuing non-proliferation and disarmament, will matter most vitally in the decades ahead.

Asia is the centre of global economic growth, military modernisation and major power competition. Demand for nuclear power is projected to rise massively across much of the region. China and India are on track to be two of the world's three biggest economies with matching diplomatic influence. Regional institutions are developing quickly. The region has only just begun to recognise and to wield its diplomatic weight globally.

Asia is the region where the use or the threat of use of nuclear weapons remains especially conceivable, notably in a US-China confrontation. And nuclear dangers remain possible in an India-Pakistan context, even if, based on current trends, a repeat of the 2002 tensions is unlikely any time soon.

So that is the regional context of these remarks: East Asia plus India, which is becoming very much integrated into East Asian trade and diplomatic structures. This region has much to lose in an era of nuclear dangers, much to gain from ameliorating these dangers before they worsen, and much scope to make efforts to this end.

Now to the substantive issue: how Asia can lead in the delegitimisation of nuclear weapons, as a stepping stone to nuclear non-proliferation and disarmament.

As we have heard very clearly these past two days, an emphasis on possession of nuclear weapons is only part of the picture: nuclear postures and doctrines are important as incentives or disincentives to disarmament.

There is an urgent global need for reducing the reliance on, and the salience of the geo-political value of nuclear weapons; in other words, achieving their delegitimisation as tools of statecraft and security policy.

The Asian experience is worth close study in this regard: most countries in the region have renounced the pursuit of nuclear weapons; and even the nuclear-armed powers, India and China, have relatively restrained nuclear postures and doctrines, various forms of no first use, low alert and negative security assurances.

The experience of delegitimising and then prohibiting chemical weapons provides an illuminating analogy: it was a regional (in this case European) norm of non-use as a stepping stone to abolition.

This norm of non-use began a long time ago, when the horrors of chemical weapons use in World War I became a major motivation for the Geneva Protocol of 1925. Even then, it had caveats: states agreed not to use CW [or indeed biological weapons (BW)] in war, but did not give up their rights to possess them, and some retained a right of retaliation in kind against their use. Basically, it was a Euro-centric no first use pact.

The record between then and the 1993 conclusion of the CWC was mixed: the Geneva Protocol remained the most significant legal instrument against the use of CW for 68 years. There was not widespread use of CW in warfare, so a taboo on use did, to some degree, take hold. But use did sometimes occur, along with many allegations of use. And, of course, the most blatant violations were Iraq's repeated use against Iranian troops and even its own Kurdish nationals in the 1980s, episodes which helped propel the CWC negotiations. But the regional norm in Europe held and in time it spread globally.

Asia, today, has the opportunity on the nuclear front to show the leadership that Europe arguably showed then but lacks today, when, for instance, the North Atlantic Treaty Organisation (NATO) is under a first use nuclear umbrella.

How can the relatively restrained nuclear weapons order now prevailing in Asia (specifically East Asia and India) be put to effective use as a building block of nuclear weapons delegitimisation towards the ultimate goal of nuclear disarmament?

A two-fold challenge and opportunity presents itself: to find ways to maintain, and strengthen and lock in a regional norm of nuclear restraint; and to consolidate and mobilise regional opinion to influence global processes in the direction of nuclear restraint, non-proliferation and disarmament.

There are opportunities here for countries such as India to creatively combine the arms control and Asian strands of their diplomacy to be prime movers in advancing these goals.

Indeed, this is an exceptional opportunity for China and India to work together. If the US and USSR, despite their glaring rivalry and differences, could coordinate their diplomacy to bring to the world the nuclear Non-Proliferation Treaty (NPT), why cannot these two new centres of wealth and power, China and India, coordinate their diplomacy to reduce nuclear dangers for all?

One way forward would be to encourage a regional leaders' dialogue on nuclear security. This might begin as a series of bilateral consultations among some key countries – perhaps India, China, Japan, Australia and Indonesia – leading to a multilateral leaders' discussion at one of the region's formal inter-state bodies, and preferably, given its membership and mandate, the East Asia Summit. This forum and its footprint have a singular congruence with nuclear restraint: of its 16 member states, only two, China and India, have nuclear weapons, and they retain relatively restrained postures.

The idea is at its core a simple one: a discussion among regional leaders, aimed at identifying, testing and expanding the common ground among their nations' interests and national thinking on nuclear security issues. This dialogue could be advanced – and ideally transformed into common positions and perhaps even some cooperation action – with varying degrees of ambition, depending on what the regional diplomatic market could bear.

The endpoint of the discussion would not be predetermined or restricted by cautious national bureaucratic positions. A challenge as critical as nuclear security deserves direct leaders-level consideration. And a primary purpose of the East Asia Summit is "open and spontaneous leaders-led discussions on strategic issues of peace and stability in our region and in the world."[1]

There should be scope in such a forum to craft an agreed declaration by regional leaders setting out principles for nuclear security. I think there is a sleeping regional consensus here awaiting mobilisation.

Non-proliferation would certainly need to feature in such a declaration, including affirmations of commitments to prevent the unsafeguarded transfer of nuclear weapons-related materials and knowledge, and to control the spread of proliferation-sensitive technology in nuclear energy programmes.

But a bolder approach should also be considered. The leaders' statement might even be an attempt to agree on the need for a restrained and stable nuclear order – precisely the kind of order the region needs if it is to prosper and to manage strategic competition involving its rising powers. This could draw on some of the conclusions of the Canberra Commission and other such reports, in particular affirming that the only acceptable role for nuclear weapons is to deter other nuclear weapons in the context of efforts to reduce nuclear arsenals.

There might be potential to specify the need for a regional order based on assurances by nuclear-armed states that they would not be the first to use nuclear weapons, that they would under no circumstances use nuclear weapons against non-nuclear states, and that their nuclear weapons are not on high alert.

A united position identifying the powers of an increasingly wealthy and influential Asia as advocate of nuclear restraint and nuclear disarmament would be too important to ignore. In addition to its regional confidence-building impact, it could have global effects. It could help to cut across the anachronistic diplomatic allegiances which have traditionally split global disarmament diplomacy into Western and Non-Aligned blocs and obstructed agreement accordingly. It might add to normative pressures on nuclear-armed powers beyond the region to reconsider changing their postures and doctrines. In particular, it could be a way of contributing fresh thinking and impetus to the 2010 NPT Review Conference process. Given this, the timing of an Asian leaders meeting would need to be within the next 18 months. East Asia Summit meetings are already scheduled for late 2008 and late 2009.

India, Australia and others could work to ensure that either or both of these sessions were expanded to involve substantial discussions on nuclear issues.

The idea of an Asian dialogue in pursuit of nuclear security, restraint and disarmament raises many obvious criticisms and questions: for instance, should the United States and Russia be involved directly or engaged in parallel? Would India take part in a discussion on nuclear doctrine that did not involve Pakistan? Would China do so without the United States? What about North Korea? Why focus on doctrines, which are changeable and, therefore, not necessarily to be trusted anyway? How to address perceptions that a regional restraint regime might simply lock in existing nuclear asymmetries (such as China's nuclear weapons inferiority to the United States, or India's to China)?

Asian conversation about the future of nuclear weapons without the US in the room will focus minds in Washington, or Moscow, like not much else will. And if these powers hold concerns about exclusion from the discussion, there can always be – indeed, there should be – parallel talks with Washington or Moscow about reducing reliance on nuclear arms.

Discussions on nuclear security in an Asian setting might require the region's nuclear-armed countries to consider, and explain, how they are contributing to global disarmament, beyond merely waiting for the United States and Russia to take the first steps. There might be a corresponding pressure on Australia, Japan and South Korea, as non-nuclear US allies in the region, to explain how their acceptance of the protection of a nuclear-armed ally is consistent with their advocacy of nuclear restraint and disarmament.

Such questions and sensitivities would inevitably influence the content of the leaders' discussions and their potential outcomes. But this is no reason not to start. The very complications and dangers of the Asian strategic environment make the need for leadership-level engagement on these issues all the greater.

Of course, the principles (and potentially eventual legal commitments) of an Asian nuclear initiative might not be specifically as outlined here. They could be in some ways more ambitious: for example, involving regional confidence-building measures (CBMs), a ban on certain classes of nuclear weapons or missiles, or commitments to certain positions vis-à-vis global treaties such as the Fissile Material Cut-Off Treaty (FMCT). Or they might be more modest depending on what the regional diplomatic market could bear at the time. No first use of nuclear weapons might, for instance, need to be considered alongside alternative formulations such as no first use of weapons of mass destruction (WMD).

The sort of regional process I am suggesting will be awkward for a number of countries not strictly part of East Asia but with interests in the region, particularly the US, Russia and Pakistan, for as long as they stick to their current nuclear doctrines and postures. That would be precisely the purpose: to identify that the new Asian mainstream is about nuclear restraint. Indeed, part of the regional dialogue process would be to build a front to lobby nuclear-armed states outside the treaty zone – including allies – to adopt principles of nuclear restraint. Ultimately, the US and Russia will need to be major players in the conversation about new and restrained road rules for the nuclear order in Asia. Pakistan's link with the region is more limited: its nuclear arsenal and first use posture are challenges for the whole international community to manage.

That still leaves North Korea, which is a difficult case on which thankfully the region is already mobilised; which should not be brought fully into the regional mainstream while it has nuclear arms; but which, in the end, might welcome a regional order containing assurances that it would not be attacked with nuclear weapons.

Like the European CW restraint regime developed from the 1920s onwards, an East Asian order built upon principles of nuclear restraint would be a building block towards global non-proliferation and disarmament, as well as a CBM in its own right.

But it would have limits, as does the CW analogy I have been drawing. The motives and diplomatic rationales behind the negotiation of the CWC were different to those which are likely to underpin efforts towards the abolition of nuclear weapons. CW are not status weapons. And their continued possession of nuclear weapons ensured that great powers were relatively comfortable with negotiating the CWC.

The diplomacy of nuclear disarmament requires a 'building block' approach: for a long time to come, the main efforts need to be about putting the bricks in place, elements such as verification, confidence-building and regional restraint.

And here, my contention is that Asia's regional multilateral forums have much to contribute; they can be used to advance regional norms of restraint and nuclear weapons delegitimisation that can in turn influence global trends, provided that Asia's great and growing diplomatic influence is properly unified and deployed.

Can Asia lead in nuclear restraint, and thus in non-proliferation and disarmament? Of course it can, provided that the leaders of the region's powers are willing to start the conversation. ∎

NOTES

1. Chairman's statement at the second East Asia Summit, Cebu, Philippines, January 15, 2007, http://www.aseansec.org/19302.htm.

Rajiv Gandhi's Vision of an NWFW in Today's Context

☐ **K. SUBRAHMANYAM**

RAJIV GANDHI'S PLEA IN 1988

Prime Minister Rajiv Gandhi presented his action plan for a "nuclear weapon free world" (NWFW) to the Special Session of the United Nations General Assembly on Disarmament on June 9, 1988. He appealed to the international community, "Nuclear war will not mean the death of a hundred million people. Or even a thousand million. It will mean the extinction of four thousand million, the end of life as we know it on our planet, earth. We come to the United Nations to seek your support. We seek your support to put a stop to this madness."

This appeal was cited and invoked in the article in the *Wall Street Journal* on January 4, 2007, which launched the initiative of four US statesmen, George Schultz, William Perry, Henry Kissinger and Sam Nunn on "A World Free of Nuclear Weapons." At the time Rajiv Gandhi made his appeal, the world had 61,000 odd nuclear weapons and it has since then come down to 20,000 in the last two decades. This is mostly due to the end of the Cold War and the ideological antagonism in a bipolar world yielding place to the birth of a polycentric balance of power system. The present international system has been described by US Secretary of State Condoleezza Rice in the following words. "For the first time since the peace of Westphalia in 1648, the prospect of violent conflict between great powers is becoming ever more unthinkable. Major states are increasingly competing in peace and not preparing for war." Russia is today a member of the G-8 group of nations which attempts to coordinate the macro-economic policies of the world. China, India, Brazil and South Africa are special invitees to that economic summit. Though the context in which Rajiv Gandhi made his appeal for nuclear disarmament has changed radically, yet his words are invoked today by those very statesmen who had contributed significantly to the build-up of global nuclear arsenals to the level of 65,000 warheads they reached in 1985 and influential sections of the international strategic community are today pleading for moving towards the goal of nuclear disarmament.

SCHULTZ-KISSINGER-PERRY-NUNN INITIATIVE 2007-08

Rajiv Gandhi said in his address, "Nuclear deterrence is the ultimate expression of the philosophy of terrorism, holding humanity hostage to the presumed security needs of a few." Twenty years later, Schultz and his colleagues state, "Nuclear weapons were essential to maintaining international security during the Cold War because they were a means of deterrence. The end of the Cold War made the doctrine of Soviet-American mutual deterrence obsolete. Deterrence continues to be a relevant consideration for many states with regard to threats from other states. But reliance on nuclear weapons for this purpose is becoming increasingly hazardous and decreasingly effective." They further add, "North Korea's recent nuclear test and Iran's refusal to stop the program to enrich uranium – potentially to weapons grade – highlight the fact that the world is now on the precipice of a new and dangerous nuclear era. Most alarmingly, the likelihood that non-state terrorists will get their hands on nuclear weaponry is increasing. In today's war, waged on the world order by terrorists, nuclear weapons are the ultimate means of mass destruction. And non-state terrorist groups with nuclear weapons are conceptually outside the bounds of a deterrent strategy and present difficult new security challenges." It is the perception of this new threat that seems to be driving the new campaigns to eliminate nuclear weapons from the arsenals of the world.

TODAY'S SITUATION

Apart from the end of the Cold War, emergence of a polycentric balance of power and globalisation of the economy, there have been other changes in the global order since Rajiv Gandhi delivered his appeal. At that time, Rajiv Gandhi warned the world, "Left to ourselves, we would not want to touch nuclear weapons. But when tactical considerations, in the passing play of great power rivalries are allowed to take precedence over the imperative of nuclear non-proliferation, with what leeway are we left?" This was a clear reference to the Chinese proliferation to Pakistan and the permissiveness of the US in that respect in the context of the Afghanistan War. He also offered restraint in respect of states that were capable of crossing the nuclear weapons threshold (meaning thereby India) if the nuclear weapon states adopted his action plan. His pleas were ignored and he was forced to conclude he had no leeway left and, consequently, he ordered the assembly of Indian nuclear weapons in March 1989.

There is a lesson here for those who are campaigning for a world without nuclear weapons on their own parochial logic without taking into account the logic of others. In terms of strategic logic, those who built arsenals of 65,000 weapons and are now dismantling those weapons at great costs to themselves and at high risks of those materials being diverted or falling into the hands of non-state actors, cannot command optimum credibility elsewhere in the world. What kind of strategic logic persuaded the five nuclear weapon nations and the rest of the nuclear Non-Proliferation Treaty (NPT) community to legitimise possession of nuclear weapons in the hands of five nuclear weapon powers by extending the NPT of 25 years duration indefinitely and unconditionally? This was done after the end of the Cold War when the consideration of nuclear deterrence could not have played as salient a role in the calculations of the strategic community as it did during the era of the Cold War.

Surely an international regime which permits legitimacy of nuclear weapons for five nations and denies it to the rest of the international community cannot be expected to be stable or command universal observance. The strategy of deterrence preached for decades made some of the signatories of the NPT who were threatened by externally induced regime change to consider acquisition of nuclear weapons to deter that eventuality. When Rajiv Gandhi pleaded for elimination of nuclear weapons, he argued that comprehensive global security must rest on a new, more just, more honourable world order. Surely, the world order which legitimises the possession of nuclear weapons in the hands of a few nations and denies them to the rest is not a just or honourable one.

THE ACTION PLAN 1988

Rajiv Gandhi proposed in 1988 the following steps in his Action Plan:

(i) A binding commitment by all nations to eliminate nuclear weapons in stages by the year 2010 at the latest.

(ii) All nuclear weapon states must participate in the process of nuclear disarmament. All other countries must also be part of the process.

(iii) To demonstrate good faith and build the required confidence, there must be tangible progress at each stage towards the common goal.

(iv) Changes are required in doctrines, policies and institutions to sustain a world free of nuclear weapons. Negotiations should be undertaken to establish a Comprehensive Global Security System under the aegis of the UN.

(v) All production of nuclear weapons and weapons grade fissile materials should stop. There should be a Comprehensive Test Ban Treaty (CTBT).

(vi) An international convention outlawing the threat or use of nuclear weapons should be negotiated.

(vii) Negotiations should commence to replace the NPT expiring in 1995 and a new treaty should give legal effect to the binding commitment of nuclear weapon states to eliminate nuclear weapons by the year 2010 at the latest.

PROPOSALS OF THE FOUR US STATESMEN

Twenty years later, George Schultz and his colleagues have put forward their proposals to move towards a nuclear weapon free world. The steps they have suggested are:

(i) Extension of key provisions of the Strategic Arms Reduction Treaty (START) of 1991.

(ii) Taking steps to increase the warning and decision times for the launch of all nuclear armed ballistic missiles, thereby reducing risks of accidental or unauthorised attacks.

(iii) Discarding any existing operational plans for massive attacks that still remain from the Cold War days.

(iv) Negotiations toward developing cooperative multilateral ballistic missile defence and early warning systems as proposed by Presidents Bush and Putin at the 2002 Moscow Summit.

(v) Acceleration of work to provide the highest possible standards of security for nuclear weapons as well as for nuclear materials everywhere in the world to prevent terrorists from acquiring the nuclear bomb.

(vi) A dialogue within the North Atlantic Treaty Organisation (NATO) and with Russia on consolidating the nuclear weapons designed for forward deployment to enhance their security and as a first step toward careful accounting for them and their eventual elimination.

(vii) Strengthening the means of monitoring compliance with the nuclear Non-Proliferation Treaty (NPT) as a counter to the global spread of advanced technologies.

(viii) Adopting a process for bringing the CTBT into effect which would strength the NPT.

These are steps within the framework of arms control but they fall far short of the goal of a nuclear weapon free world. The four statesmen suggest that there should also be an agreement to undertake further substantive reduction in US and Russian nuclear forces beyond those recorded in the US-Russian SORT (Strategic Offensive Reduction Treaty). They hope that as the reductions proceed, other nuclear weapon nations would become involved. They also advocate a verifiable treaty to prevent nations from producing nuclear materials for weapons. They want an international consensus to be built to deter or when required, to respond to secret attempts by countries to break out of agreements.

In order to facilitate progress on these measures, they urge a clear statement of the ultimate goal of a world free of nuclear weapons. They confess that the goal is like the top of a very tall mountain whose top they are unable to see at present and, therefore, there must be a charting of a course to higher ground where the mountain top becomes more visible.

In a world where 90 per cent of the nuclear arsenal is with two powers, the US and Russia, it is natural for the US statesmen to be heavily preoccupied with mostly arms control reductions and further steps within that framework, and keep the nuclear weapon free world as a very distant goal. It is quite obvious that they do not envisage any meaningful commitment to a nuclear weapon free world within the foreseeable future. In fact, the steps they have listed out highlight that nuclear weapons will stay for quite some time to come. In such circumstances, one wonders whether their advocacy of a nuclear weapon free world will be any more meaningful than the commitment of the five nuclear weapon powers in Article VI of the NPT to negotiate nuclear disarmament. The only point of discontinuity is that those who used to argue that a nuclear weapon free world was a utopian concept, nuclear weapons could not be disinvented and only arms control was realistic, have now started talking about the need to spell out the ultimate goal of a nuclear weapon free world. Objectively speaking, this is, no doubt progress but not very significant progress.

FIGHTABILITY OF A NUCLEAR WAR

There are already comments from devotees of the arms control approach and NPT in the West that the vision of a nuclear weapon free world is unrealistic. Rajiv Gandhi, in his address, five years before the treaty on the elimination of chemical weapons was signed, asked why the nuclear weapons could not be eliminated when there was a convention to eliminate biological weapons and negotiations were on for eliminating chemical weapons. Underlying all the mental blocks against the elimination of nuclear weapons appears to be a lingering perception that the nuclear weapons are usable weapons of war. On this issue, writing in *Foreign Policy*, May/June 2005, former US Defence Secretary Robert McNamara said,

I have worked on issues relating to US and NATO nuclear strategy and war plans for more than 40 years. During that time, I have never seen a piece of paper that outlined a plan for the United States or NATO to initiate the use of nuclear weapons with any benefit to the United States or NATO. I have made this statement in front of audiences, including NATO defence ministers and senior military leaders many times. No one has ever refuted it... I reached these conclusions very soon after becoming Secretary of Defence. Although I believe Presidents John F. Kennedy and Lyndon Johnson shared my view, it was impossible for any of us to make such statements publicly because they were totally contrary to established NATO policy.

Yet, in spite of this conclusion reached early in his career as defence secretary, which, according to him, was shared by Presidents Kennedy and Johnson, the US nuclear arsenal rose from 24,000 in 1961 to 29,000 in 1968 when he left office.

On the US command and control over nuclear weapons Dr Bruce Blair, writing in his "Nuclear Column" on February 16, 2004, quoted General Lee Butler, the former Strategic Force commander (1991-94):

Part of the insidiousness of the evolution of this system ... is the unfortunate fact that, whatever might have been intended by the-policy makers (who, incidentally had very little insight into the mechanisms that underpinned the simple words that floated onto a blank page at the level of the White House), in reality, at the operational level, the requirements of deterrence proved impracticable....The consequence was a move in practice to a system structured to drive the president invariably toward a decision to launch under attack... Launch under attack means that you believe you have incontrovertible proof that warheads actually are on the way... Our policy was premised on being able to accept the first wave of attacks. We never said publicly that we were committed to launch on warning or launch under attack. Yet, at the operational level, it was never accepted that if the presidential decision went to a certain tick of the clock, we would lose a major portion of our forces... Notwithstanding the intention of deterrence as it is expressed at the policy level — as it is declared and written down — at the level of the operations, those intentions got turned on their head, as the people who are responsible for actually devising the war plan faced the dilemmas and blind alleys of concrete practice. Those mattered absolutely to the people who had to sit down and try to frame the detailed guidance to exact destruction of 80 per cent of the adversary's nuclear forces. When they realised that they could not, in fact, assure those levels of damage if the President chose to ride out an attack, what then did they do? They built a construct that powerfully biased the President's decision process toward launch before the arrival of the first enemy warhead.

These two authoritative pronouncements raise the question of whether a nuclear war between two nuclear-armed nations, at comparable levels of capability, is fightable and winnable. President Reagan and General Secretary Gorbachev had declared in Geneva in 1985 that a nuclear war could not be won and should not be initiated. While there have been many commissions to further the cause of non-proliferation, there has never been a commission of retired Strategic Force commanders to examine whether a nuclear war was fightable in a militarily meaningful sense.

DELEGITIMISING NUCLEAR WEAPONS

This is a crucial issue in any initiative to eliminate nuclear weapons and reach a nuclear weapon free world. It was the realisation following the use of chemical weapons by both sides in World War I that they were not battle-winning instrumentalities that led to the Geneva Protocol of 1925 which established "a no first use" agreement among countries capable of producing chemical weapons. Both the realisation of futility of use of chemical weapons to win a war and a sense of deterrence since both sides possessed them in abundant quantities ensured that they were not used in World War II. The chemical weapons were used only in asymmetric situations when the aggressor had them and the victim did not and also in a permissive international environment when the international community callously chose to look away. Such were the cases of use of chemical weapons in the Ethiopian War of the Thirties and Saddam Hussein's war on Iran in the Eighties. Therefore, a professional military conclusion on the fightability and winnability of a nuclear war will help to clear the fog of nuclear strategy as popularised in the traditional nuclear theology.

No weapon considered usable as a war-winning instrumentality and deemed legitimate by the international community is likely to get eliminated. Most of the non-proliferationists have been found to be reluctant to address the issue of the legitimacy of nuclear weapons and to go into the question of fightability of a nuclear war. Rajiv Gandhi proposed 20 years ago that all nuclear weapons be leached of legitimacy by negotiating an international convention which outlawed the threat or use of such weapons. He argued that such a convention will reinforce the process of nuclear disarmament.

NUCLEAR THEOLOGY AND ITS IMPACT

In today's globalised world, with fast growing interdependence among nations, the doctrine of deterrence serves to reinforce the terrorism of *jehadis* who too justify their use of violence or threat of violence involving mass destruction to achieve their political and religious purposes in the same way as the major nuclear powers did during the Cold War. The basic acceptance of terrorism as the instrumentality of politics is the core of deterrence theology, as Rajiv pointed out two decades ago. In the case of nuclear deterrence, as has been practised among nations, the threat of use was adequate to project deterrence since the weapon has been used twice with devastating effect. The *jehadis* may be thinking of similar demonstration effect in order to exercise deterrence against nation-states they believe are present in their area against the wishes of the people of a particular faith. The doctrine of nuclear deterrence has come under severe criticism by many, including very distinguished senior military commanders who had been in charge of nuclear weapons and forces. In the post-Cold War era, some non-democratic states which are concerned about the externally induced regime changes, have acquired or attempt to acquire nuclear weapons to deter such moves against them on the basis of the logic the nuclear theologians have been advocating. They are able to take advantage of the laxity in export controls of advanced nuclear industrial countries, especially in Western Europe. In a sense, the doctrine of deterrence operates in such cases against major nuclear powers, deterring them and often compelling them to make concessions. Arms control treaties like the NPT can be no guarantees against such countries breaking out since major powers have abrogated other arms control treaties such as the Anti- Ballistic Missile (ABM) Treaty and have not honoured their obligations under Article VI of the NPT. North Korea has demonstrated that a country can walk out of the NPT.

In the aftermath of 9/11 and the discovery of the decades long international nuclear technology and materials black market, certain steps have been initiated outside the NPT to reinforce the non-proliferation regime. Notable among them is the Proliferation Security Initiative (PSI). However, it remains to be seen how far these steps will be effective in the light of continued legitimacy of nuclear weapons, espousal of deterrence doctrines and laxity of export controls in some advanced industrial countries and past permissiveness of proliferation by the non-proliferation community itself. These shortcomings in upholding the provisions of the NPT have been repeatedly highlighted in the inability of successive Review Conferences to come out with positive findings on compliance with Articles I and II of the NPT.

Among the means available for countering proliferation of nuclear technology, materials and manufacturing equipment to the non-state terrorist actors, the most effective is, as advocated by late Professor Joseph Rotblat, the Nobel Peace prize winner, societal verification. But that will be possible only if the nuclear weapons are delegitimised. So long as the weapons are legitimate for some nations, there will be strong nationalist pressures to be permissive of surreptitious acquisition of nuclear weapons and technologies among peoples of other countries, especially countries not part of any major international security system and which face major security problems, some of which are fallouts of the Cold War. Most of the non-proliferation debates have been led in fora where the apparent beneficiaries of deterrent strategies of the Cold War dominate. Their bias is towards sustaining the status quo which they, according to some influential commentators, wrongly believe preserved the peace during Cold War era.

GLOBAL APPROACH TO NUCLEAR ISSUE: INDIAN PROPOSALS

Rajiv Gandhi was perhaps one of the few leaders who addressed the issue of elimination of nuclear weapons from the global instead of parochial perspective. He was prepared to forego Indian nuclear weaponisation if his plan was accepted. But his pleas and warnings went unheeded. Therefore, twenty years later, we find that his appeal and Action Plan are still relevant with new possible threats of weapons and nuclear materials falling into the hands of non-state terrorist actors and some more threatened regimes breaking out of their NPT obligations to deter attempts directed against them for regime change. Therefore, it is not surprising that the latest suggestions of the Indian delegate in the Conference on Disarmament in Geneva reflect the essentials of the Rajiv Gandhi plan. India has also urged the appointment of a special coordinator to assist in carrying out consultations on specific measures or a set of measures that have the potential of commanding consensus which can form the basis of a mandate for a possible ad hoc committee on nuclear disarmament. The seven steps India has suggested are:

1. Reaffirmation of the unequivocal commitment of all nuclear weapon states to the goal of complete elimination of nuclear weapons.

2. Reduction of the salience of nuclear weapons in the security doctrines.

3. Adoption of measures by nuclear weapon states to reduce nuclear danger, including the risks of accidental nuclear war, de-alerting of nuclear weapons to prevent unintentional and accidental use of weapons.

4. Negotiation of a global agreement among nuclear weapon states on "no first use" of nuclear weapons.

5. Negotiation of a universal and legally binding agreement on non-use of nuclear weapons against non-nuclear weapon states.

6. Negotiation of a convention on the complete prohibition of the use or threat of use of nuclear weapons.

7. Negotiation of a nuclear weapons convention prohibiting the development production, stockpiling and use of nuclear weapons and on their destruction, leading to the global non-discriminatory and verifiable elimination of nuclear weapons within a specified time- frame.

It may be noted that the initiative of the four US statesmen is step 3 in the above list. They have problems in seeing the top of the high mountain that constitutes the goal of disarmament because they do not start with the first steps proposed nor take them further to the logical conclusion. What is needed today is the political will among the leaderships of the nine nuclear weapon states to commit themselves that they will not use the nuclear weapon as an instrument of war, they will not rely excessively on the unproven and unprovable effectiveness of nuclear deterrence, they will delegitimise the use and threat of use of nuclear weapons and start negotiating a non-discriminatory treaty to achieve a verifiable elimination of nuclear weapons on the model of the Chemical Weapons Convention. India went nuclear after a clear warning by Rajiv Gandhi to the international community. However, India has already declared its commitment to "no first use," and rejection of the nuclear war-fighting strategy. India's nuclear weapons are solely meant to deter the use and threat of use of nuclear weapons against it since other nuclear weapon nations still adhere to the doubtful rationality of nuclear deterrence and, therefore, it has become compelling to deter those who subscribe to nuclear deterrence doctrines and resort to them in international relations.

Let the leaders of nuclear weapon countries be bold enough to study and establish whether a nuclear war is fightable or winnable in a meaningful military sense. In my view, it is not. If that can be publicly established by a commission of former Strategic Force commanders from the nine nuclear weapon states, we would have taken a giant step towards delegitimisation of nuclear weapons and their elimination. There may be people who may ask whether it may not lead to a dubious conclusion on the lines of the judgement of the International Court of Justice (ICJ) on the legality of use of nuclear weapons. One cannot rule out that possibility. Even in that case, the resulting situation will not be worse than what it is today, as has happened in the case of the ICJ judgement. If the findings were to be that nuclear weapons will be militarily meaningfully usable, then that will generate pressures on the countries that want to keep nuclear weapons to themselves and deny them to others and shake the foundation of the NPT. However, there are demands from some of the nuclear weapon powers themselves that the salience of nuclear weapons in the security doctrines should be reduced, indicating that there is reason to be hopeful that a military professional study on the fightability of a nuclear war need not produce any surprisingly damaging findings.

It is not uncommon in history for statesmen and military commanders continuing to be conditioned by obsolete thinking and venturing into disastrous adventures. World Wars I and II, the Vietnam War, the Afghanistan War in the Eighties, the Iraq-Iran War and the ongoing Iraq War were the results of such confused and obsolete thinking. But those disasters took place in a world driven by animosity. Today, there is a world in which politico-military establishments of leading countries are in a balance of power, non-antagonistic situation and, therefore, in a position to study jointly an issue like the fightability of nuclear war.

One can understand the reluctance of those who have used nuclear weapons and missiles as currency of power in the second half of the 20th century to be involved in such a study. There is growing realisation that in the 21st century, knowledge will be the currency of power and no longer missiles and nuclear weapons in an international system of global interdependence among the major powers (six of them nuclear). Yet, since the leaderships of the five vintage nuclear weapon powers may be conditioned by the conventional nuclear strategic dogma, the initiative for this study must come from the sixth nuclear power which is recognised today as an emerging balancer – India. It would be an appropriate tribute to Rajiv Gandhi and his legacy if India were to set up an international commission of former nuclear force commanders to study the fightability of a nuclear war. Rajiv Gandhi's vision of a nuclear weapon free world is as relevant today as it was 20 years ago. It is time that India did something more concrete than engaging itself in the conventional diplomacy on the nuclear issue. ∎

Some Concluding Thoughts

☐ **MANPREET SETHI**

To discover new oceans, one must have the courage to lose sight of the shore.

OVER THE last decade or so, the security challenges dominating the nuclear discourse have involved concerns about the violation of the nuclear Non-Proliferation Treaty (NPT) by member states, the presence of sophisticated networks for illicit marketing of nuclear material and technology, and the threat of nuclear terrorism. International energies have accordingly been directed towards either creating new military doctrines of preemption and counter-proliferation to deal with proliferation, and/or to buttress national or international technology control and denial regimes to check illicit nuclear trade. However, these measures can only partially address contemporary nuclear challenges.

Nations have proven time and again that they will find ways and resources to acquire nuclear weapons if they consider it in their national interest to do so. Therefore, enforcing non-proliferation calls for more than negative controls. It also calls for positive incentives such as improvement in the security environment and a revocation of the policy that allows nuclear weapons with some but not with others. Common sense would point out that this is simply not sustainable, and that is how it has turned out over the last six decades that nuclear weapons have been around.

Today, the expansion of nuclear energy programmes in order to meet growing energy needs of an increasing population in developing countries, and to mitigate availability as well as environmental concerns of thermal power sources, poses a new challenge of reconciling nuclear power with

---

(This chapter is based on personal reflections on some of the issues raised at the Conference.)

proliferation. The use of nuclear power entails an inherent duality of purpose and if the world desires to exploit the peaceful use of the atom, as it must for human development, without having to constantly fear for proliferation, then it is inevitable that a non-discriminatory framework be crystallised to address these issues. A world that has renounced the military use of the atom would logically be the most conducive for the extensive, peaceful use of this potent energy source.

However, the attainment of a secure and stable nuclear weapon free world naturally calls for a large number of measures. These would have to range from fostering greater mutual trust in inter-state relations, a greater transparency in defence matters, and a greater range and depth of arms control efforts. Given the complexities involved and the overlap of issues that would practically require an intense churning of international relations, it appears far easier, and less risky, to keep trudging along the known course – to stay within the familiar perimeters of nuclear deterrence. Consequently, for most decision-makers everywhere, the idea of universal nuclear disarmament is indulgently humoured as a fine long-term aspiration – something to be kept in mind, but not yet to be taken in practical terms as a sensible policy goal.

Unfortunately, such a mindset refuses to accept that times have changed. Given the security challenges that arise from the presence of nuclear weapons today – the threat of deliberate or inadvertent nuclear exchange, spread of nuclear weapons to non-nuclear states, and the spectre of terrorism, given the ease of access to nuclear materials and technologies – nuclear weapons are no longer 'providers' of security (if they ever were), but a liability that demands more security to keep them on leash. Isn't it time then to think afresh? In fact, a new wave of thought that can alter existing mindsets on the utility of nuclear weapons has to be the essential starting point for any change that can come about in the physical state of nuclear weapons.

History proves that institutions whose abolition was once thought to be unthinkable have broken down only under the pressure of changed belief systems. It may be recalled that when a few had started demanding the abolition of slavery, there were several who argued against it on the grounds that the economic system and society would not be able to survive the drastic change. At the time, on December 1, 1862, President Abraham Lincoln in his famous message to the US Congress had said, "The dogmas of the quiet past are inadequate to the stormy present... As our case is new, so we must think anew and act anew. We must disenthrall ourselves."

This advice has come in useful whenever radical transformations have taken place. It is most relevant today in the case of nuclear weapons. We have to disenthrall ourselves from their hold and if anything can do that, it is the force of new ideas. In 1988, India's Prime Minister Rajiv Gandhi attempted just this by presenting a bold and comprehensive Action Plan before the international community. This document clearly spelt out a set of measures to be undertaken in three phases spread over 22 years so that the security perceptions of the nuclear and non-nuclear states could be equally addressed. Nuclear disarmament constituted the centrepiece of each stage of the plan but it was to be scaffolded by collateral measures such as steps to preclude the development of new weapon systems based on emerging technologies, reducing conventional arms and forces to minimum levels required for defensive purposes and the conduct of international relations on the principles of non-violence and coexistence. Had this plan been accepted in 1988, the world would have been only some months short of bidding farewell to nuclear weapons today!

However, the nuclear free and non-violent world anchored in cooperative security that Rajiv Gandhi visualised proved to be far ahead of the times in which he lived. Ironically, in fact, even as he was working on the draft of the Action Plan, blatant proliferation raged in the region and his address received an indifferent response at the UN. On his return to New Delhi from New York, Rajiv Gandhi was provided intelligence reports that clearly pointed towards Pakistan having acquired a nuclear weapons capability from China. He and his country were jolted into accepting the regional reality and the international mood on the issue.

Considerably chastised by the existing state of things in the region and beyond, India felt compelled to acquire nuclear deterrence to deal with the challenges to its national security. However, India has never abandoned its desire for an a nuclear weapon free world (NWFW), even though disillusioned with the response the idea evokes elsewhere. This innate belief in the NWFW stems from the hardcore realistic understanding of how India's national security would be significantly enhanced if it did not have to contend with nuclear weapons. The presence of nuclear weapons in the region complicates India's security. Even as the US and the international community worry about the possibility of nuclear terrorism, India has been living with the reality of terrorism – of the level of a sub-conventional conflict – cheekily mounted by Pakistan from behind the skirts of nuclear weapons. For India, an NWFW is a security imperative.

Not surprisingly, therefore, in his inaugural address at the conference, Prime Minister Dr Manmohan Singh stressed India's commitment to nuclear disarmament that is global, universal and non-discriminatory, and which could be realised through a collective approach anchored in a universal partnership supported by non-governmental communities and public opinion. He suggested several steps that could be taken in this regard but emphasised that India did not consider them either exhaustive or hold them in any rigid hierarchy. He expressed India's openness to explore and support any and all measures that may contribute to achieving the goal.

It merits some consideration as to what possibly could be the conditions that might be conducive for an NWFW. In the following paragraphs, let me attempt to identify four possible circumstances that could promote the creation of an environment in which nuclear disarmament would seem less undesirable and more feasible. The availability of these fortuitous conditions would automatically mean a change in not only our nuclear belief systems but also in our concepts of security, thereby making the attainment of a nuclear weapon free world more plausible.

Firstly, common sense would have us believe that no nuclear weapon state (NWS), nor any other state desirous of becoming one, would want to renounce its nuclear capability/ambitions, until it is convinced that some other more reliable means providing for its security is firmly in place. This sense of security, in turn, could be instilled in two ways: one, by effectuating an intrinsic change in the international security environment through the conclusion of a variety of confidence building measures (CBMs) that promise a global order premised on cooperative rather than competitive security; and/or secondly, by attaining a clear superiority with conventional weapons. At one level, these two prerequisites of a more secure international order appear to be mutually exclusive because amassing conventional weapons could rather vitiate the international security environment. However, it needs to be remembered that in this case the conventional weapons would actually be replacing weapons of mass destruction (WMD) and the likely spurt in their accumulation might prove to be transitional, occurring only soon after the actual abolition of nuclear weapons and more to serve a psychological, rather than a military purpose. Moreover, if the surrender of nuclear weapons is accompanied by, or is a result of, more profound changes in the international security regime, then the entire prevailing inter-state paradigm where every nation enhances its own security by building offensive and defensive military capabilities that decrease the perceived security of others would become redundant. The recent financial crisis has exposed the vulnerabilities of the global economic network and the close-knit nature of contemporary inter-state relations; the nuclear interdependence of action and reaction should not be ignored.

Secondly, states can be expected to renounce their nuclear option only when enough technical expertise is available to ensure adequate verification of the commitments undertaken by the states to abandon their nuclear arsenals. One of the major reasons why nuclear disarmament has been considered unfeasible until now is because of the lack of an adequate mechanism, both institutional as well as technological, to oversee the process of disarmament. The inability to decide on what to do with the dismantled warheads and how to ensure their safe and secure storage has proved to be as much a handicap as the lack of mutual trust among nations.

Historical experience illustrates that belief systems change only when it becomes technically possible to prove something. For centuries, man continued to believe that the sun revolves around the earth. Those who disputed this belief were ridiculed or persecuted. Copernicus in his lifetime could not find acceptance for his theory. The belief changed only after the invention of the telescope and other astronomical instruments that could technically prove that the sun was actually at the centre of the universe.

Applying the same logic to the issue of nuclear disarmament, it may be said that advancements in the field of surveillance and monitoring technologies coupled with the ongoing research on how to deal with the nuclear materials obtained from dismantled warheads would help to change the belief that nuclear disarmament is practically non-doable. And this is possible today. The experience of the Cooperative Threat Reduction programme between the US and Russia indicates the possibilities that exist in verification, dismantlement and safe use of weapons material for peaceful uses. It is estimated that today, 20 per cent of all electricity in the US comes from nuclear power plants; 50 per cent of all nuclear fuel used in these plants comes from dismantled Russian weapons. Interestingly, therefore, one out of every ten bulbs in the US is today powered by material that was once in Soviet nuclear warheads!!

Thirdly, by a rather perverse logic, movement towards an NWFW could be propelled if there was an accumulation of enough risks from the continued possession of these weapons. If the dangers surmount and keep on doing so, then states might see sense in doing away with these weapons. Some of these risks, in palpable proportions, are already becoming evident. For instance, we are living in times when ethnic conflicts, religious fundamentalism, proliferation of a terrorist ethos, "loose nukes", illegal trafficking in nuclear weapons or materials, etc are on an upward swing. The reasons for conflict are also increasing as competition becomes as fierce for scarce renewable resources (water and fertile land), as for non-renewable resources (oil and minerals), even as the planet becomes dangerously overpopulated. At the same time, the modern international system is experiencing a technology push in which more and more and increasingly sophisticated technology is available from an ever growing number of suppliers. This erodes the effectiveness of technology export controls and heightens the risk of nuclear proliferation. In fact, even more ominous than the flow of components and technology may prove to be the transformation of nuclear experts (designers, engineers, etc.) from an elite core confined in space to a floating population in search of jobs in a resurgent nuclear industry in several parts of the world.

In an age where interest in nuclear energy is legitimately surfacing as a safe, secure and environmentally sustainable source of electricity, the international community would find it far easier to manage the growth of the nuclear industry for peaceful purposes if the presence of nuclear weapons could be done away with. Of course, this would not wipe away the knowledge of making weapons or completely obviate the possibility of some nations cheating on their commitment not to acquire nuclear weapons. But a universal, international treaty that abolishes nuclear weapons would make the task of handling a few violators far easier than having to deal with an open ended situation.

Fourthly, and most importantly, the prospects of nuclear disarmament would be brighter, if, based on the above three factors, enough pressure was to be mounted on the governments to make politically binding commitments to this effect. Fortunately, some strains of this are beginning to surface again after they last died down in the end 1990s. At the individual level, security analysts and even former military men have begun to question the rationale behind the continuance of nuclear weapons. Cold War practitioners such as George Shultz, Henry Kissinger, William Perry, and Sam Nunn have expressed themselves in favour of disarmament. In two articles that resounded across the world in January 2007 and 2008, the four gentlemen who were responsible for crafting airtight nuclear deterrence doctrines are today conceding the fallacy of that logic. Also of import are the myriad attempts being made elsewhere – a study to examine the verification challenges of disarmament in the UK; a Commission set up by Australia and Japan to study the prospects of disarmament; and the Global Zero initiative led by Bruce Blair, a command and control expert for American weapons only some years ago.

However, the moot point is whether these initiatives will outlast the 2010 NPT Review Conference (RevCon). Or are they targeted at showcasing a benign nuclear face of the NWS before the non-nuclear international community at the forthcoming RevCon in order to coerce them into swallowing the sugarcoated pill of diluting Article IV of the NPT that would restrict access to the full fuel cycle technologies? If the various strands being woven globally do not result in coming together to form the disarmament fabric, it will be another historic opportunity lost. And even more importantly and sadly, the credibility of future initiatives towards a nuclear weapon free world would suffer since the sceptics would gain more ammunition to argue that nuclear disarmament is practically impossible to achieve.

Therefore, sincerity of purpose is as important as the action itself. While one cannot set a date as to when an NWFW can be realised, it can, however, be said with some certainty that if mindsets change, it could come about quickly. Ideas have the power to snowball into a force capable of bringing about a profound change. It may not be forgotten that every revolution in history has taken its leaders by surprise. In fact, as long as the old order exists, its control seems total and impregnable. But, as cracks develop, its fragility is quickly exposed.

What is needed is the presence of a critical mass of decision-makers who are willing to take the risk of thinking afresh and trying out a new state security system. In the ultimate analysis, political will holds the key to the decision; technology, accumulation of dangers or rise of public opinion can only be the facilitators. As indicated at the start of this chapter, new oceans can only be discovered if we are ready to move deep where the known shores begin to fade away and new possibilities begin to emerge. With nuclear weapons too, the comfort zone of nuclear deterrence will have to be abandoned in order to visualise a new, non-nuclear world. ■

A World Free of Nuclear Weapons

☐ RAJIV GANDHI

WE ARE approaching the close of the twentieth century. It has been the most bloodstained century in history. Fifty-eight million perished in two World Wars. Forty million more have died in other conflicts. In the last nine decades, the ravenous machines of war have devoured nearly one hundred million people. The appetite of these monstrous machines grows on what they feed. Nuclear war will not mean the death of a hundred million people. Or even a thousand million. It will mean the extinction of four thousand million: the end of life as we know it on our planet Earth. We come to the United Nations to seek your support. We seek your support to put a stop to this madness.

Humanity is at a crossroads. One road will take us like lemmings to our own suicide. That is the path indicated by doctrines of nuclear deterrence, deriving from traditional concepts of the balance of power. The other road will give us another chance. That is the path signposted by the doctrine of peaceful coexistence, deriving from the imperative values of non-violence, tolerance and compassion.

In consequence of doctrines of deterrence, international relations have been gravely militarised. Astronomical sums are being invested in ways of dealing with death. Ever new means of destruction continue to be invented. The best of our scientific talent and the bulk of our technological resources are devoted to maintaining and upgrading this awesome ability to obliterate ourselves. A culture of armaments and threats and violence has become pervasive.

For a hundred years after the Congress of Vienna, Europe knew an uncertain peace based on a balance of power. When that balance was tilted – or more accurately, when that balance was perceived to have been tilted – Europe was plunged into an orgy of destruction, the like of which had never been known before and which spread to engulf much of the world. The unsettled disputes of the First World War led to the Second.

Humankind survived because, by today's standards, the power to destroy, which was then available was a limited power. We now have what we did not then have: the power to ensure the genocide of the human race. Technology has now rendered obsolete the calculations of war and peace on which were constructed the always dubious theories of the balance of power.

It is a dangerous delusion to believe that nuclear weapons have brought us peace. It is true that in the past four decades, parts of the world have experienced an absence of war. But a mere absence of war is not a durable peace. The balance of nuclear terror rests on the retention and augmentation of nuclear armouries. There can be no ironclad guarantee against the use of weapons of mass destruction. They have been used in the past. They could be used in the future. And, in this nuclear age, the insane logic of mutually assured destruction will ensure that nothing survives, that none lives to tell the tale, that there is no one left to understand what went wrong and why. Peace which rests on the search for a parity of power is a precarious peace. If we can understand what went wrong with such attempts in the past, we may yet be able to escape the catastrophe presaged by doctrines of nuclear deterrence.

There is a further problem with deterrence. The doctrine is based on the assumption that international relations are frozen on a permanently hostile basis. Deterrence needs an enemy, even if one has to be invented. Nuclear deterrence is the ultimate expression of the philosophy of terrorism: holding humanity hostage to the presumed security needs of a few.

There are those who argue that since the consequences of nuclear war are widely known and well- understood, nuclear war just cannot happen. Neither experience nor logic can sustain such dangerous complacency. History is full of miscalculations. Perceptions are often totally at variance with reality. A madman's fantasy could unleash the end. An accident could trigger a chain reaction, which inexorably leads to doom. Indeed, the advance of technology has so reduced the time for decisions that, once activated, computers programmed for Armageddon preempt human intervention and all hope of survival. There is, therefore, no comfort in the claim of the proponents of nuclear deterrence that everyone can be saved by ensuring that in the event of conflict, everyone will surely die.

The champions of nuclear deterrence argue that nuclear weapons have been invented and, therefore, cannot be eliminated. We do not agree. We have an international convention eliminating biological weapons by prohibiting their use in war. We are working on similarly eliminating chemical weapons. There is no reason on principle why nuclear weapons too cannot be so eliminated. All it requires is the affirmation of certain basic moral values and the assertion of the required political will, underpinned by treaties and institutions, which insure against nuclear delinquency.

The past few years have seen the emergence of a new danger: the extension of the nuclear arms race into outer space. The ambition of creating impenetrable defences against nuclear weapons has merely escalated the arms race and complicated the process of disarmament. This has happened in spite of the grave doubts expressed by leading scientists about its very feasibility. Even the attempt to build a partial shield against nuclear missiles increases the risk of nuclear war. History shows that there is no shield that has not been penetrated by a superior weapon, nor any weapon for which a superior shield has not been found. Societies get caught in a multiple helix of escalation in chasing this chimera, expending vast resources for an illusory security while incurring the risk of certain extinction.

The new weapons being developed for defence against nuclear weapons are part of a much wider qualitative arms race. The development of the so-called "third generation nuclear weapons" has opened up ominous prospects of their being used for selective and discriminate military operations. There is nothing more dangerous than the illusion of limited nuclear war. It desensitises inhibitions about the use of nuclear weapons. That could lead, in next to no time, to the outbreak of full-fledged nuclear war.

There are no technological solutions to the problems of world security. Security can only come from our asserting effective political control over this self-propelled technological arms race. We cannot accept the logic that a few nations have the right to pursue their security by threatening the survival of humankind. It is not only those who live by the nuclear sword who, by design or default, shall one day perish by it. All humanity will perish.

Nor is it acceptable that those who possess nuclear weapons are freed of all controls while those without nuclear weapons are policed against their production. History is full of such prejudices paraded as iron laws: that men are superior to women; that the white races are superior to the coloured; that colonialism is a civilising mission; that those who possess nuclear weapons are responsible powers and those who do not are not.

Alas, nuclear weapons are not the only weapons of mass destruction. New knowledge is being generated in the life sciences. Military applications of these developments could rapidly undermine the existing convention against the military use of biological weapons. The ambit of our concern must extend to all means of mass annihilation.

New technologies have also dramatically expanded the scope and intensity of conventional warfare. The physical destruction, which can be carried out by full-scale conventional war, would be enormous, far exceeding anything known in the past. Even if humankind is spared the agony of a nuclear winter, civilisation and civic life as we know it would be irretrievably disrupted. The range, precision and lethality of conventional weapons is being vastly increased. Some of these weapons are moving from being 'smart' to becoming 'intelligent'. Such diabolical technologies generate their own pressures for early use, thus, increasing the risk of the outbreak of war. Most of these technologies are at the command of the military blocs. This immensely increases their capacity for interference, intervention and coercive diplomacy.

Those of us who do not belong to the military blocs would much rather stay out of the race. We do not want to accumulate arms. We do not want to augment our capacity to kill. But the system, like a whirlpool, sucks us into its vortex. We are compelled to divert resources from development to defence to respond to the arsenals, which are constructed as a sideshow to great power rivalries. As the nature and sophistication of threat to our security increase, we are forced to incur huge expenditure on raising the threshold of our defences.

There is another danger that is even worse. Left to ourselves, we would not want to touch nuclear weapons. But when tactical considerations, in the passing play of great power rivalries, are allowed to take precedence over the imperative of nuclear non-proliferation, with what leeway are we left?

Even the mightiest military powers realise that they cannot continue the present arms race without inviting economic calamity. The continuing arms race has imposed a great burden on national economies and the global economy. It is no longer only the developing countries that are using disarmament to channel resources to development. Even the richest are beginning to realise that they cannot afford the current levels of the military burden they have imposed upon themselves. A genuine process of disarmament, leading to a substantial reduction in military expenditure, is bound to promote the prosperity of all nations of the globe. Disarmament accompanied by coexistence will open up opportunities for all countries, whatever their socio-economic systems, whatever their levels of development.

The technological revolutions of our century have created unparalleled wealth. They have endowed the fortunate with high levels of mass consumption and widespread social welfare. In fact, there is plenty for everyone, provided distribution is made more equitable. Yet, the possibility of fulfilling the basic needs of nutrition and shelter, education and health remains beyond the reach of vast millions of people in the developing world because resources which could give fulfillment in life are preempted for death.

The root causes of global insecurity reach far below the calculus of military parity. They are related to the instability spawned by widespread poverty, squalor, hunger, disease and illiteracy. They are connected to the degradation of the environment. They are enmeshed in the inequity and injustice of the present world order. The effort to promote security for all must be underpinned by the effort to promote opportunity for all and equitable access to achievement. Comprehensive global security must rest on a new, more just, more honourable world order.

When the General Assembly met here last in Special Session to consider questions of disarmament, the outlook was grim. The new Cold War had been revived with full force. A new programme of nuclear armament had been set in motion. As a result, during the years that followed, fear and suspicion cast a long shadow over all disarmament negotiations. Humankind was approaching the precipice of nuclear disaster.

Today, there is a new hope for survival and for peace. There is a perceptible movement away from the precipice. Dialogue has been resumed. Trust is in the air.

How has this transformation occurred? We pay tribute to the sagacity of the American and Soviet leaderships. They have seen the folly of nuclear escalation. They have started tracing the outlines of a pattern of disarmament. At the same time, we must recognise the role of countless enlightened men and women all over the world, citizens of the non-nuclear weapon states as much as of the nuclear weapon states. With courage, dedication and perseverance they kept the candle burning in the enveloping darkness. The Six-Nation Initiative voiced the hopes and aspirations of these many millions. At a time when relations between the two major nuclear weapon states dipped to their nadir, the Six Nations – Argentina, Greece, India, Mexico, Sweden and Tanzania – refocussed world attention on the imperative of nuclear disarmament. The appeal of May 1984, issued by Indira Gandhi, Olof Palme and their colleagues, struck a responsive chord. Negotiations stalled for years began inching forwards. The process begun in Geneva has led to Reykjavik, Washington and Moscow.

We have all welcomed the ratification of the Intermediate Nuclear Forces (INF) Treaty concluded between General Secretary Gorbachev and President Reagan. It is an important step in the right direction. Its great value lies in its bold departure from nuclear arms limitation to nuclear disarmament. We hope there will be agreement soon to reduce strategic nuclear arsenals by 50 per cent. The process should be carried forward to the total elimination of nuclear weapons. Only then will we be able to look back and say that the INF Treaty was a truly historic beginning. India believes it is possible for the human race to survive the second millennium. India believes it is also possible to ensure peace, security and survival into the third millennium and beyond. The way lies through concerted action. We urge the international community to immediately undertake negotiations with a view to adopting a time-bound Action Plan to usher in a world order free of nuclear weapons and rooted in non-violence.

We have submitted such an Action Plan to this Special Session on Disarmament of the United Nations General Assembly. Our plan calls upon the international community to negotiate a binding commitment to general and complete disarmament. This commitment must be total. It must be without reservation.

The heart of our Action Plan is the elimination of all nuclear weapons, in three stages, over the next twenty-two years, beginning now. We put this plan to the United Nations as a programme to be launched at once.

While nuclear disarmament constitutes the centerpiece of each stage of the plan, this is buttressed by collateral and other measures to further the process of disarmament. We have made proposals for banning other weapons of mass destruction. We have suggested steps for precluding the development of new weapon systems based on emerging technologies. We have addressed ourselves to the task of reducing conventional arms and forces to the minimum levels required for defensive purposes. We have outlined ideas for the conduct of international relations in a world free of nuclear weapons.

The essential features of the Action Plan are:

First, there should be a binding commitment by all nations to eliminating nuclear weapons in stages, by the year 2010 at the latest.

Second, all nuclear weapon states must participate in the process of nuclear disarmament. All other countries must also be part of the process.

Third, to demonstrate good faith and build the required confidence, there must be tangible progress at each stage towards the common goal.

Fourth, changes are required in doctrines, policies and institutions to sustain a world free of nuclear weapons. Negotiations should be undertaken to establish a Comprehensive Global Security System under the aegis of the United Nations.

We propose simultaneous negotiations on a series of integrally related measures. But we do recognise the need for flexibility in the staging of some of these measures.

In Stage-I, the INF Treaty must be followed by a fifty per cent cut in Soviet and US strategic arsenals. All production of nuclear weapons and weapons grade fissionable material must cease immediately. A moratorium on the testing of nuclear weapons must be undertaken with immediate effect to set the stage for negotiations on a Comprehensive Test Ban Treaty.

It is already widely accepted that a nuclear war cannot be won and must not be fought. Yet, the right is reserved to resort to nuclear war. This is incompatible with a binding commitment to the elimination of nuclear weapons. Therefore, we propose that all nuclear weapons be leached of legitimacy by negotiating an international convention which outlaws the threat or use of such weapons. Such a convention will reinforce the process of nuclear disarmament.

Corresponding to such a commitment by the nuclear weapon states, those nations, which are capable of crossing the nuclear weapons threshold, must solemnly undertake to restrain themselves. This must be accompanied by strict measures to end all covert and overt assistance to those seeking to acquire nuclear weapons.

We propose that negotiations must commence in the first stage itself for a new treaty to replace the NPT, which expires in 1995. This new treaty should give legal effect to the binding commitment of nuclear weapon states to eliminate all nuclear weapons by the year 2010 and of all the non-nuclear weapon states to not cross the nuclear weapons threshold.

International law already bans the use of biological weapons. Similar action must be taken to ban chemical and radiological weapons.

The international community has unanimously recognised outer space as the common heritage of mankind. We must expand international cooperation in the peaceful use of outer space. The essential prerequisite for this is that outer space be kept free of all weapons. Instead, there are plans of developing, testing and deploying a space weapons system. The nuclear arms race cannot be ended and reversed without a moratorium on such activity. It should be followed by an agreement to forestall the militarisation of outer space. This is also an indispensable condition for attaining the goal of comprehensive global security based on a non-violent world order free of nuclear weapons.

The very momentum of developments in military technology is dragging the arms race out of political control. The race cannot be restrained without restraining the development of such technology. We need a system which fosters technological development but interdicts its application to military purposes. The arms control approach has focussed on the quantitative growth of arsenals. The disarmament approach must devise arrangements for controlling the continuous qualitative upgradation of nuclear and conventional weapons. To achieve this purpose, the essential requirement is increased transparency in research and development in frontier technologies with potential military applications. This requires a systematic monitoring of such developments, an assessment of their implications for international security, and widespread dissemination of the information obtained. There is also need for greater international cooperation in research into new and emerging technologies for these technologies to open on new vistas of human achievement. Here, let us recall the vision of an open world voiced by one of the most remarkable scientists of our time, Niels Bohr. In his Open Letter to the United Nations on June 9, 1950, thirty-eight years ago today, he said:

The very fact that knowledge itself is a basis for civilisation points directly to openness as the way to overcome the present crisis.

By the closing years of the century, there must be a single integrated multilateral verification system to ensure that no new nuclear weapons are produced anywhere in the world. Such a system would also help in verifying compliance with the collateral and other disarmament measures envisaged in the Action Plan. It would serve as an early warning system to guard against violations of solemn international treaties and conventions.

Beyond a point, nuclear disarmament itself would depend upon progress in the reduction of conventional armaments and forces. Therefore, a key task before the international community is to ensure security at lower levels of conventional defence. Reductions must, of course, begin in areas where the bulk of the world's conventional arms and forces are concentrated. However, other countries should also join the process without much delay. This requires a basic restructuring of armed forces to serve defensive purposes only. Our objective should be nothing less than a general reduction of conventional arms across the globe to levels dictated by minimum needs of defence. The process would require a substantial reduction in offensive military capabilities as well as confidence building measures to preclude surprise attacks. The United Nations needs to evolve by consensus a new strategic doctrine of non-provocative defence.

The plan for radical and comprehensive disarmament must be pursued along with efforts to create a new system of comprehensive global security. The components of such a system must be mutually supportive. Participation in it must be universal.

The structure of such a system should be firmly based on non-violence. When we eliminate nuclear weapons and reduce conventional forces to minimum defensive levels, the establishment of a non-violent world order is the only way of not relapsing into the irrationalities of the past. It is the only way of precluding the recommencement of an armaments spiral. Non-violence in international relations cannot be considered an Utopian goal. It is the only available basis for civilised survival, for the maintenance of peace through peaceful coexistence, for a new, just, equitable and democratic world order. As Mahatma Gandhi said in the aftermath of the first use of nuclear weapons:

> The moral to be legitimately drawn from the supreme tragedy of the bomb is that it will not be destroyed by counter-bombs, even as violence cannot be destroyed by counter-violence. Mankind has to get out of violence only through non-violence.

The new structure of international relations must be based on respect for various ideologies, on the right to pursue different socio-economic systems, and the celebration of diversity. Happily, this is already beginning to happen. Post-war bipolarity is giving way to a growing realisation of the need for coexistence. The high rhetoric of the system of military alliances is gradually yielding to the viewpoint of the Non-Aligned Movement.

Non-alignment is founded on the desire of nations for freedom of action. It stands for national independence and self-reliance. Non-alignment is a refusal to be drawn into the barren rivalries and dangerous confrontations of others. It is an affirmation of the need for self-confident cooperation among all countries, irrespective of differences in social and economic systems. Non-alignment is synonymous with peaceful coexistence. As Jawaharlal Nehru said:

> The alternative to co-existence is co-destruction.

Therefore, the new structure of international relations to sustain a world beyond nuclear weapons will have to be based on the principles of coexistence, the non-use of force, non-intervention in the internal affairs of other countries, and the right of every state to pursue its own path of development. These principles are enshrined in the Charter of the United Nations, but they have been frequently violated. We must apply our minds to bringing about the institutional changes required to ensure their observance. The strengthening of the United Nations system is essential for comprehensive global security. We must resurrect the original vision of the United Nations. We must bring the United Nations Organisation in line with the requirements of the New World Order.

The battle for peace, disarmament and development must be waged both within this assembly and outside by the peoples of the world. This battle should be waged in cooperation with scientists, strategic thinkers and leaders of peace movements who have repeatedly demonstrated their commitment to those ideals. We, therefore, seek their cooperation in securing the commitment of all nations and all peoples to the goal of a non-violent world order free of nuclear weapons.

The ultimate power to bring about change rests with the people. It is not the power of weapons or economic strength, which will determine the shape of the world beyond nuclear weapons. That will be determined in the minds and the hearts of thinking men and women around the world. For, as the Dhammapada of the Buddha teaches us:

> Our life is shaped by our mind; we become what we think. Suffering follows
> an evil thought as the wheels of a cart follow the oxen that draw it. Joy follows
> a pure thought like a shadow that never leaves. For hatred can never put an
> end to hatred; love alone can. This is the unalterable law. ∎

Action Plan for Ushering in a Nuclear Weapon Free and Non-Violent World Order

1. Humanity stands at a crossroads of history. The world has lived too long under the sentence of extinction. Nuclear weapons threaten to annihilate human civilisation and all that humankind has built through millennia of labour and toil. Nuclear-weapon states and non-nuclear-weapon states alike are threatened by such a holocaust. It is imperative that nuclear weapons be eliminated. The recently signed INF Treaty between the United States and the Soviet Union is a first major step in this direction. This process must be taken to its logical conclusion by ridding the world of nuclear weapons. The time has also come to consider seriously the changes in doctrines, in policies, in attitudes, and in the institutions required to usher in and manage a nuclear - weapon- free and non-violent world. Peace must be predicated on a basis other than the assurance of global destruction. We need a world order based on non-violence and peaceful coexistence. We need international institutions that will nurture such a world order.

2. We call upon the international community to urgently negotiate a binding commitment to an action plan for ushering in a non-violent world free of nuclear weapons. We suggest the following action plan as a basis for such negotiations:

2.1.	**STAGE I** (duration: 6 years, from 1988 to 1994)
2.1.a.	**Nuclear disarmament:**
2.1.a.i.	Elimination of all Soviet and United States land-based medium and shorter range missiles (500 to 5,500 km) in accordance with the INF Treaty.
2.1.a.ii.	Agreement on a 50 percent cut in Soviet and United States strategic arsenals (with ranges above 5, 500 km)
2.1.a.iii.	Agreement on a phased elimination by the year 2000 A.D. of United States and Soviet short range battlefield and air-launched nuclear weapons.
2.1.a.iv.	Cessation of the production of nuclear weapons by all nuclear

	weapon states.
2.1.a.v.	Cessation of production of weapon-grade fissionable material by all nuclear weapon states.
2.1.a.vi.	Moratorium on the testing of nuclear weapons.
2.1.a.vii.	Commencement and conclusion of negotiations of a comprehensive test-ban treaty.
2.1.b.	**Measures collateral to nuclear disarmament:**
2.1.b.i.	Conclusion of a convention to outlaw the use and threat of use of nuclear weapons pending their elimination,
2.1.b.ii.	Declaration by the United States and the Soviet Union that the fissile material released under the INF Treaty would be utilised for peaceful purposes only and accordingly be subjected to supervision by the International Atomic Energy Agency.
2.1.b.iii.	Declaration by all nuclear weapon states of their stockpiles of nuclear weapons and weapon-grade fissionable material.
2.1.b.iv.	Cessation of direct or indirect transfer to other states of nuclear weapons, delivery systems, and weapon-grade fissionable material.
2.1.b.v.	Non-nuclear weapon powers to undertake not to cross the threshold into the acquisition of nuclear weapons.
2.1.b.vi.	Initiation of multilateral negotiations to be concluded by 1995 for a new treaty eliminating all nuclear weapons by the year 2010. This treaty would replace the Non-Proliferation Treaty which ends in 1995.
2.1.c.	**Other weapons of mass destruction:**
2.1.c.i.	Conclusion of a treaty banning chemical weapons.
2.1.c.ii.	Conclusion of a treaty banning radiological weapons.
2.1.d.	**Conventional forces:**
2.1.d.i.	Substantial reduction of NATO and Warsaw Pact conventional forces, especially offensive forces, and of weapons system in Europe from the Atlantic to the Urals.
2.1.d.ii.	Multilateral discussions in the Conference on Disarmament or in the United Nations on military doctrines with a view to working towards the goal of a purely defensive orientation for the armed forces of the world. The discussions would include measures to prevent surprise attacks.
2.1.e.	**Space weapon systems:**
2.1.e.i.	A moratorium on the testing and deployment of all space

weapon systems.

2.1.e.ii.   Expansion of international cooperation in the peaceful uses of outer space.

2.1.f   **Control and management of the arms race based on new technologies:**

2.1.f.i   Arrangements for monitoring and assessing new technologies, which have military applications, as well as forecasting their implications for international security.

2.1.fii   For research in frontier areas of technology where there are potential military applications, new technology projects and technological missions should be undertaken under the auspices of the United Nations in order to direct them exclusively to civilian sectors.

2.1.f.iii   Commencement of work, under the aegis of the United Nations, for the formulation of guidelines to be observed by governments in respect of new technologies with potential military applications.

2.1.f.iv   Commencement of negotiations for banning technological missions designed to develop new weapon systems and means of warfare.

2.1.g   **Verification:**

2.1.g.i   Acceptance in principle of the need to establish an integrated multilateral verification system under the aegis of the United Nations as an integral part of a strengthened multilateral framework required to ensure peace and security during the process of disarmament as well as in a nuclear weapon free world.

2.2   **STAGE II (duration: 6 years, from 1995 to 2000)**

2.2.a   **Nuclear Disarmament:**

2.2.a.i   Completion of Stage I reductions by the United States and the Soviet Union and the induction of all other nuclear weapon states into the process of nuclear disarmament.

2.2.a.ii   Elimination of all medium and short range, sea-based, land-based and air-launched nuclear missiles by all nuclear weapon states.

2.2.a.iii   Elimination of all tactical battlefield nuclear weapons (land, sea and air) by all nuclear weapon states.

2.2.a.iv   Entry into force of the comprehensive test ban treaty (CTBT).

2.2.b	**Measures collateral to nuclear disarmament**
2.2.b.i	Negotiations on withdrawal of strategic nuclear weapons deployed beyond national boundaries.
2.2.b.ii	Completion of the ratification and entry into force of the convention prohibiting the use and threat of use of nuclear weapons.
2.2.b.iii	Conclusion of the new treaty eliminating all nuclear weapons by the year 2010 to replace the non-proliferation treaty (NPT).
2.2.c	**Space weapons:**
2.2.c.i	Agreement within a multilateral framework on banning the testing, development, deployment and storage of all space weapons.
2.2.d.	**Conventional forces:**
2.2.d.i.	Further reduction of NATO and Warsaw Pact conventional forces to minimum defensive levels.
2.2.d.ii.	Negotiations under the Conference on Disarmament on global conventional arms reduction.
2.2.d.iii.	Removal of all military forces and bases from foreign territories.
2.2.e.	**New and emerging technologies:**
2.2.e.i.	Completion of negotiations on banning technological missions aimed at the development of new weapon systems.
2.2.e.ii.	Completion of negotiations on guidelines in respect of new technologies with potential military applications.
2.2.f.	**Comprehensive global security system:**
2.2.f.i.	Negotiations on and establishment of a comprehensive global security system to sustain a world without nuclear weapons. This would include institutional steps to ensure the effective implementation of the provisions of the Charter of the United Nations relating to the non-use of force, the peaceful settlement of disputes, and the right of every state to pursue its own path of development.
2.2.f.ii.	Arrangements for the release of resources through disarmament for development purposes.
2.2.f.iii.	Elimination of non-military threats to security by such measures as the establishment of a just and equitable international economic order.

2.2.f.iv.	The strengthening of the United Nations system and related multilateral forums.
2.2.f.v.	The commencement of negotiations for the establishment of an integrated multilateral verification system under the United Nations.
2.3.	**STAGE III** (duration: 10 years, from 2001 to 2010)
2.3.a.	Elimination of all nuclear weapons from the world.
2.3.b.	Establishment of a single integrated multilateral comprehensive verification system which *inter alia*, ensures that no nuclear weapons are produced.
2.3.c.	Reduction of all conventional forces to minimum defensive levels.
2.3.d.	Effective implementation of arrangements to preclude the emergence of a new arms race.
2.3.e.	Universal adherence to the comprehensive global security system.
3.1.	There has been a historically unprecedented militarisation of international relations during the last four decades. This has not only enhanced the danger of nuclear war but also militated against the emergence of the structure of peace, progress and stability envisaged in the Charter of the United Nations.
3.2.	To end this dangerous militarisation of international relations, we must build a structure firmly based on non-violence. It is only in a non-violent democratic world that sovereignty of nations and the dignity of the individual can be ensured. It is only in a non-violent world that the intellectual and spiritual potential of humankind can be fully realised.
3.3.	The prospect of a world free from nuclear weapons should spur us to start building a structure of international security in keeping with the fundamental changes that are taking place in the world political, economic and security environment.
3.4.	In a shrinking and interdependent world, such a structure has to be comprehensive, its components supportive of each other, and participation in it universal.
3.5.	A world order crafted out of outmoded concepts of the balance of power, of dominance by power blocs, of spheres of influence, and of special rights and privileges for a select group of nations is an unacceptable anachronism. It is out of tune with the democratic temper of our age.

3.6. The new structure of international relations has to be based on scrupulous adherence to the principles of peaceful coexistence and the Charter of the United Nations. It is necessary to evolve stronger and more binding mechanisms for the settlement of disputes, regional and international. The diversity among nations must be recognised and respected. The right of each nation to choose its own socio-economic system must be assured.

3.7. Concomitant changes will be called for in the international economic order. The interdependence of all the economies of the world makes for a symbiotic relationship between development in the South and stability and growth in the North. In a just and equitable order, access to technology and resources, on fair and reasonable terms will be assured. The gap between the rich and the poor nations will be bridged.

A World Free of Nuclear Weapons*

□ **GEORGE P. SHULTZ, WILLIAM J. PERRY, HENRY A. KISSINGER** AND **SAM NUNN**

PART I

NUCLEAR WEAPONS today present tremendous dangers, but also an historic opportunity. US leadership will be required to take the world to the next stage – to a solid consensus for reversing reliance on nuclear weapons globally as a vital contribution to preventing their proliferation into potentially dangerous hands, and ultimately ending them as a threat to the world.

Nuclear weapons were essential to maintaining international security during the Cold War because they were a means of deterrence. The end of the Cold War made the doctrine of mutual Soviet-American deterrence obsolete. Deterrence continues to be a relevant consideration for many states with regard to threats from other states. But reliance on nuclear weapons for this purpose is becoming increasingly hazardous and decreasingly effective.

North Korea's recent nuclear test and Iran's refusal to stop its program to enrich uranium – potentially to weapons grade – highlight the fact that the world is now on the precipice of a new and dangerous nuclear era. Most alarmingly, the likelihood that non-state terrorists will get their hands on nuclear weaponry is increasing. In today's war, waged on world order by terrorists, nuclear weapons are the ultimate means of mass devastation. And non-state terrorist groups with nuclear weapons are conceptually outside the bounds of a deterrent strategy and present difficult new security challenges.

Apart from the terrorist threat, unless urgent new actions are taken, the US soon will be compelled to enter a new nuclear era that will be more precarious, psychologically disorienting, and economically even more costly than was Cold War deterrence. It is far from certain that we can successfully replicate the old Soviet-American "mutually assured destruction" with an increasing number of potential nuclear enemies worldwide without dramatically

* *The Wall Street Journal*, January 4, 2007.

increasing the risk that nuclear weapons will be used. New nuclear states do not have the benefit of years of step-by-step safeguards put in effect during the Cold War to prevent nuclear accidents, misjudgments or unauthorized launches. The United States and the Soviet Union learned from mistakes that were less than fatal. Both countries were diligent to ensure that no nuclear weapon was used during the Cold War by design or by accident. Will new nuclear nations and the world be as fortunate in the next 50 years as we were during the Cold War?

Leaders addressed this issue in earlier times. In his "Atoms for Peace" address to the United Nations in 1953, Dwight D. Eisenhower pledged America's "determination to help solve the fearful atomic dilemma – to devote its entire heart and mind to find the way by which the miraculous inventiveness of man shall not be dedicated to his death, but consecrated to his life." John F. Kennedy, seeking to break the logjam on nuclear disarmament, said, "The world was not meant to be a prison in which man awaits his execution."

Rajiv Gandhi, addressing the UN General Assembly on June 9, 1988, appealed, "Nuclear war will not mean the death of a hundred million people. Or even a thousand million. It will mean the extinction of four thousand million: the end of life as we know it on our planet earth. We come to the United Nations to seek your support. We seek your support to put a stop to this madness."

Ronald Reagan called for the abolishment of "all nuclear weapons," which he considered to be "totally irrational, totally inhumane, good for nothing but killing, possibly destructive of life on earth and civilization." Mikhail Gorbachev shared this vision, which had also been expressed by previous American presidents.

Although Reagan and Mr. Gorbachev failed at Reykjavik to achieve the goal of an agreement to get rid of all nuclear weapons, they did succeed in turning the arms race on its head. They initiated steps leading to significant reductions in deployed long- and intermediate-range nuclear forces, including the elimination of an entire class of threatening missiles.

What will it take to rekindle the vision shared by Reagan and Mr. Gorbachev? Can a worldwide consensus be forged that defines a series of practical steps leading to major reductions in the nuclear danger? There is an urgent need to address the challenge posed by these two questions.

The Non-Proliferation Treaty (NPT) envisioned the end of all nuclear weapons. It provides (a) that states that did not possess nuclear weapons as of 1967 agree not to obtain them, and (b) that states that do possess them agree to divest themselves of these weapons over time. Every president of both parties since Richard Nixon has reaffirmed these treaty obligations, but non-nuclear weapon states have grown increasingly skeptical of the sincerity of the nuclear powers.

Strong non-proliferation efforts are under way. The Cooperative Threat Reduction program, the Global Threat Reduction Initiative, the Proliferation Security Initiative and the Additional Protocols are innovative approaches that provide powerful new tools for detecting activities that violate the NPT and endanger world security. They deserve full implementation. The negotiations on proliferation of nuclear weapons by North Korea and Iran, involving all the permanent members of the Security Council plus Germany and Japan, are crucially important. They must be energetically pursued.

But by themselves, none of these steps are adequate to the danger. Reagan and General Secretary Gorbachev aspired to accomplish more at their meeting in Reykjavik 20 years ago – the elimination of nuclear weapons altogether. Their vision shocked experts in the doctrine of nuclear deterrence, but galvanized the hopes of people around the world. The leaders of the two countries with the largest arsenals of nuclear weapons discussed the abolition of their most powerful weapons.

What should be done? Can the promise of the NPT and the possibilities envisioned at Reykjavik be brought to fruition? We believe that a major effort should be launched by the United States to produce a positive answer through concrete stages.

First and foremost is intensive work with leaders of the countries in possession of nuclear weapons to turn the goal of a world without nuclear weapons into a joint enterprise. Such a joint enterprise, by involving changes in the disposition of the states possessing nuclear weapons, would lend additional weight to efforts already under way to avoid the emergence of a nuclear-armed North Korea and Iran.

The program on which agreements should be sought would constitute a series of agreed and urgent steps that would lay the groundwork for a world free of the nuclear threat. Steps would include:

- Changing the Cold War posture of deployed nuclear weapons to increase warning time and thereby reduce the danger of an accidental or unauthorized use of a nuclear weapon.
- Continuing to reduce substantially the size of nuclear forces in all states that possess them.
- Eliminating short-range nuclear weapons designed to be forward-deployed.
- Initiating a bipartisan process with the Senate, including understandings to increase confidence and provide for periodic review, to achieve ratification of the Comprehensive Test Ban Treaty, taking advantage of recent technical

advances, and working to secure ratification by other key states.

- Providing the highest possible standards of security for all stocks of weapons, weapons-usable plutonium, and highly enriched uranium everywhere in the world.
- Getting control of the uranium enrichment process, combined with the guarantee that uranium for nuclear power reactors could be obtained at a reasonable price, first from the Nuclear Suppliers Group and then from the International Atomic Energy Agency (IAEA) or other controlled international reserves. It will also be necessary to deal with proliferation issues presented by spent fuel from reactors producing electricity.
- Halting the production of fissile material for weapons globally; phasing out the use of highly enriched uranium in civil commerce and removing weapons-usable uranium from research facilities around the world and rendering the materials safe.
- Redoubling our efforts to resolve regional confrontations and conflicts that give rise to new nuclear powers.

Achieving the goal of a world free of nuclear weapons will also require effective measures to impede or counter any nuclear-related conduct that is potentially threatening to the security of any state or peoples.

Reassertion of the vision of a world free of nuclear weapons and practical measures toward achieving that goal would be, and would be perceived as, a bold initiative consistent with America's moral heritage. The effort could have a profoundly positive impact on the security of future generations. Without the bold vision, the actions will not be perceived as fair or urgent. Without the actions, the vision will not be perceived as realistic or possible.

We endorse setting the goal of a world free of nuclear weapons and working energetically on the actions required to achieve that goal, beginning with the measures outlined above.

TOWARDS A NUCLEAR FREE WORLD
PART II*

The accelerating spread of nuclear weapons, nuclear know-how and nuclear material has brought us to a nuclear tipping point. We face a very real possibility that the deadliest weapons ever invented could fall into dangerous hands.

The steps we are taking now to address these threats are not adequate to the danger. With nuclear weapons more widely available, deterrence is decreasingly effective and increasingly hazardous.

One year ago, in an essay in this paper, we called for a global effort to

---

* *The Wall Street Journal*, January 15, 2008.

reduce reliance on nuclear weapons, to prevent their spread into potentially dangerous hands, and ultimately to end them as a threat to the world. The interest, momentum and growing political space that has been created to address these issues over the past year has been extraordinary, with strong positive responses from people all over the world.

Mikhail Gorbachev wrote in January 2007 that, as someone who signed the first treaties on real reductions in nuclear weapons, he thought it his duty to support our call for urgent action: "It is becoming clearer that nuclear weapons are no longer a means of achieving security; in fact, with every passing year they make our security more precarious."

In June, the United Kingdom's foreign secretary, Margaret Beckett, signaled her government's support, stating: "What we need is both a vision – a scenario for a world free of nuclear weapons – and action – progressive steps to reduce warhead numbers and to limit the role of nuclear weapons in security policy. These two strands are separate but they are mutually reinforcing. Both are necessary, but at the moment too weak."

We have also been encouraged by additional indications of general support for this project from other former US officials with extensive experience as secretaries of state and defense and national security advisors. These include: Madeleine Albright, Richard V. Allen, James A. Baker III, Samuel R. Berger, Zbigniew Brzezinski, Frank Carlucci, Warren Christopher, William Cohen, Lawrence Eagleburger, Melvin Laird, Anthony Lake, Robert McFarlane, Robert McNamara and Colin Powell.

Inspired by this reaction, in October 2007, we convened veterans of the past six administrations, along with a number of other experts on nuclear issues, for a conference at Stanford University's Hoover Institution. There was general agreement about the importance of the vision of a world free of nuclear weapons as a guide to our thinking about nuclear policies, and about the importance of a series of steps that will pull us back from the nuclear precipice.

The US and Russia, which possess close to 95% of the world's nuclear warheads, have a special responsibility, obligation and experience to demonstrate leadership, but other nations must join.

Some steps are already in progress, such as the ongoing reductions in the number of nuclear warheads deployed on long-range, or strategic, bombers and missiles. Other near-term steps that the US and Russia could take, beginning in 2008, can in and of themselves dramatically reduce nuclear dangers. They include:

- Extend key provisions of the Strategic Arms Reduction Treaty of 1991. Much has been learned about the vital task of verification from the application of these provisions. The treaty is scheduled to expire on Dec. 5, 2009. The key provisions of this treaty, including their essential monitoring and verification requirements, should be extended, and the further reductions agreed upon in the 2002 Moscow Treaty on Strategic Offensive Reductions should be completed as soon as possible.

- Take steps to increase the warning and decision times for the launch of all nuclear-armed ballistic missiles, thereby reducing risks of accidental or unauthorized attacks. Reliance on launch procedures that deny command authorities sufficient time to make careful and prudent decisions is unnecessary and dangerous in today's environment. Furthermore, developments in cyber-warfare pose new threats that could have disastrous consequences if the command-and-control systems of any nuclear-weapons state were compromised by mischievous or hostile hackers. Further steps could be implemented in time, as trust grows in the US-Russian relationship, by introducing mutually agreed and verified physical barriers in the command-and-control sequence.

- Discard any existing operational plans for massive attacks that still remain from the Cold War days. Interpreting deterrence as requiring mutual assured destruction (MAD) is an obsolete policy in today's world, with the US and Russia formally having declared that they are allied against terrorism and no longer perceive each other as enemies.

- Undertake negotiations toward developing cooperative multilateral ballistic-missile defense and early warning systems, as proposed by Presidents Bush and Putin at their 2002 Moscow summit meeting. This should include agreement on plans for countering missile threats to Europe, Russia and the US from the Middle East, along with completion of work to establish the Joint Data Exchange Center in Moscow. Reducing tensions over missile defense will enhance the possibility of progress on the broader range of nuclear issues so essential to our security. Failure to do so will make broader nuclear cooperation much more difficult.

- Dramatically accelerate work to provide the highest possible standards of security for nuclear weapons, as well as for nuclear materials everywhere in the world, to prevent terrorists from acquiring a nuclear bomb. There are nuclear weapons materials in more than 40 countries around the world, and there are recent reports of alleged attempts to smuggle nuclear material in Eastern Europe and the Caucasus. The US, Russia and other nations that have worked with the Nunn-Lugar programs, in cooperation with the International Atomic Energy Agency (IAEA), should play a key role in helping to implement United Nations Security Council Resolution 1540 relating to improving nuclear security – by offering teams to assist jointly any nation in meeting its obligations under this resolution to provide for appropriate, effective security of these materials.

As Gov. Arnold Schwarzenegger put it in his address at our October conference, "Mistakes are made in every other human endeavor. Why should nuclear weapons be exempt?" To underline the governor's point, on August 29-30, 2007, six cruise missiles armed with nuclear warheads were loaded on a US Air Force plane, flown across the country and unloaded. For 36 hours, no one knew where the warheads were, or even that they were missing.

- Start a dialogue, including within NATO and with Russia, on consolidating the nuclear weapons designed for forward deployment to enhance their security, and as a first step toward careful accounting for them and their eventual elimination. These smaller and more portable nuclear weapons are, given their characteristics, inviting acquisition targets for terrorist groups.

- Strengthen the means of monitoring compliance with the nuclear Non-Proliferation Treaty (NPT) as a counter to the global spread of advanced technologies. More progress in this direction is urgent, and could be achieved through requiring the application of monitoring provisions (Additional Protocols) designed by the IAEA to all signatories of the NPT.

- Adopt a process for bringing the Comprehensive Test Ban Treaty (CTBT) into effect, which would strengthen the NPT and aid international monitoring of nuclear activities. This calls for a bipartisan review, first, to examine improvements over the past decade of the international monitoring system to identify and locate explosive underground nuclear tests in violation of the CTBT; and, second, to assess the technical progress made over the past decade in maintaining high confidence in the reliability, safety and effectiveness of the nation's nuclear arsenal under a test ban. The Comprehensive Test Ban Treaty Organization is putting in place new monitoring stations to detect nuclear tests – an effort the US should urgently support even prior to ratification.

In parallel with these steps by the US and Russia, the dialogue must broaden on an international scale, including non-nuclear as well as nuclear nations.

Key subjects include turning the goal of a world without nuclear weapons into a practical enterprise among nations, by applying the necessary political will to build an international consensus on priorities. The government of Norway will sponsor a conference in February that will contribute to this process.

Another subject: Developing an international system to manage the risks of the nuclear fuel cycle. With the growing global interest in developing nuclear energy and the potential proliferation of nuclear enrichment capabilities, an international program should be created by advanced nuclear countries and a strengthened IAEA. The purpose should be to provide for reliable supplies of nuclear fuel, reserves of enriched uranium, infrastructure assistance, financing, and spent fuel management – to ensure that the means to make nuclear weapons materials isn't spread around the globe.

There should also be an agreement to undertake further substantial reductions in US and Russian nuclear forces beyond those recorded in the US-Russia Strategic Offensive Reductions Treaty. As the reductions proceed, other nuclear nations would become involved.

President Reagan's maxim of "trust but verify" should be reaffirmed. Completing a verifiable treaty to prevent nations from producing nuclear materials for weapons would contribute to a more rigorous system of accounting and security for nuclear materials.

We should also build an international consensus on ways to deter or, when required, to respond to, secret attempts by countries to break out of agreements.

Progress must be facilitated by a clear statement of our ultimate goal. Indeed, this is the only way to build the kind of international trust and broad cooperation that will be required to effectively address today's threats. Without the vision of moving toward zero, we will not find the essential cooperation required to stop our downward spiral.

In some respects, the goal of a world free of nuclear weapons is like the top of a very tall mountain. From the vantage point of our troubled world today, we can't even see the top of the mountain, and it is tempting and easy to say we can't get there from here. But the risks from continuing to go down the mountain or standing pat are too real to ignore. We must chart a course to higher ground where the mountaintop becomes more visible. ■

Statement by George P. Shultz, William J. Perry, Henry A. Kissinger and Sam Nunn on the occasion of the 20th anniversary of former Prime Minister Shri Rajiv Gandhi's June 9, 1988, speech to the United Nations General Assembly

TWENTY YEARS ago, the late Prime Minister Shri Rajiv Gandhi addressed the United Nations General Assembly. He spoke of the nuclear dangers facing our world, and the responsibility we all share to work for global nuclear disarmament.

Prime Minister Gandhi's historic address built on a strong foundation of advocacy for global nuclear disarmament by a consensus of Indian leaders spanning political parties which has persisted for many decades.

The world is now on the precipice of a new and even more dangerous nuclear era. We must reverse reliance on nuclear weapons globally as a vital contribution to preventing their spread and ultimately ending them as a threat to the world.

We looked to Prime Minister Gandhi's speech as we wrote an article in January 2007 calling for rekindling the vision of a nuclear weapon free world. We share your commitment to seeing his vision made real. We believe that the vision will be achieved by working energetically on the steps required to achieve that goal. Without the bold vision, the actions will not be perceived as fair or urgent, and without the actions, the vision will not be perceived as realistic or possible.

Dealing decisively with the nuclear threat is the greatest challenge facing the international community today. Establishing a solid consensus for reversing reliance on nuclear weapons globally and ultimately ending them as a threat to the world will be the responsibility of all states who possess these weapons and will require the cooperation of all nations, non-nuclear as well as nuclear.

We welcome India's continued leadership and commitment on this crucial issue and send our warm wishes to you on this anniversary of Prime Minister Gandhi's speech. ■

This statement was sent to the organisers ofthe Conference (CSIS & ICWA) on June 8, 2008.